Killer Content

Addison-Wesley Information Technology Series

Capers Jones, Series Editor

The information technology (IT) industry is in the public eye now more than ever before because of a number of major issues in which software technology and national policies are closely related. As the use of software expands, there is a continuing need for business and software professionals to stay current with the state of the art in software methodologies and technologies. The goal of the Addison-Wesley Information Technology Series is to cover any and all topics that affect the IT community: These books illustrate and explore how information technology can be aligned with business practices to achieve business goals and support business imperatives. Addison-Wesley has created this innovative series to empower you with the benefits of the industry experts' experience.

For more information point your browser to
http://www.awl.com/cseng/series/it/

Wayne Applehans, Alden Globe, and Greg Laugero, *Managing Knowledge: A Practical Web-Based Approach.* ISBN: 0-201-43315-X

Gregory C. Dennis and James R. Rubin, *Mission-Critical Java™ Project Management: Business Strategies, Applications, and Development.* ISBN: 0-201-32573-X

Kevin Dick, *XML: A Manager's Guide.* ISBN: 0-201-43335-4

Jill Dyché, *e-Data: Turning Data into Information with Data Warehousing.* ISBN: 0-201-65780-5

Capers Jones, *Software Assessments, Benchmarks, and Best Practices.* ISBN: 0-201-48542-7

Capers Jones, *The Year 2000 Software Problem: Quantifying the Costs and Assessing the Consequences.* ISBN: 0-201-30964-5

Ravi Kalakota and Marcia Robinson, *e-Business: Roadmap for Success.* ISBN: 0-201-60480-9

David Linthicum, *Enterprise Application Integration.* ISBN: 0-201-61583-5

Sergio Lozinsky, *Enterprise-Wide Software Solutions: Integration Strategies and Practices.* ISBN: 0-201-30971-8

Patrick O'Beirne, *Managing the Euro in Information Systems: Strategies for Successful Changeover.* ISBN: 0-201-60482-5

Mai-lan Tomsen, *Killer Content: Strategies for Web Content and E-Commerce.* ISBN: 0-201-65786-4

Bill Wiley, *Essential System Requirements: A Practical Guide to Event-Driven Methods.* ISBN: 0-201-61606-8

Bill Zoellick, *Web Engagement: Connecting to Customers in e-Business.* ISBN: 0-201-65766-X

Killer Content

Strategies for Web Content and E-Commerce

Mai-lan Tomsen

An imprint of Addison Wesley Longman, Inc.

Reading, Massachusetts • Harlow, England • Menlo Park, California • Berkeley, California • Don Mills, Ontario • Sydney • Bonn • Amsterdam • Tokyo • Mexico City

The publisher offers discounts on this book when ordered in quantity for special sales. For more information, please contact:

Corporate, Government, and Special Sales
Addison Wesley Longman, Inc.
One Jacob Way
Reading, Massachusetts 01867

Library of Congress Cataloging-in-Publication Data

Tomsen, Mai-lan
Killer content : strategies for Web content and E-commerce / Mai-lan Tomsen.
p. cm—(The Addison-Wesley information technology series)
Includes bibliographical references and index.
ISBN 0-201-65786-4
1. Electronic commerce. 2. World Wide Web. 3. Internet (Computer network) I. Title: Strategies for Web content and E-commerce. II. Title. III. Series.

HF5548.32.T65 2000
658.8'4--dc21 00-025663

ISBN 0201657864

Text printed on recycled and acid-free paper.

1 2 3 4 5 6 7 8 9 10 –MA– 04 03 02 01 00

First printing, April 2000

To my husband,
Mark; my parents,
Peter and Kim;
and my sister, Kim-I

Contents

CHAPTER SIX DESIGNING WEB INFORMATION STRUCTURE 135

CHAPTER SEVEN FACILITATING WEB SITE PROCESSES 159

List of Sidebars

Individual Site Value Exchange

Strategies to Enhance Users' Experiences

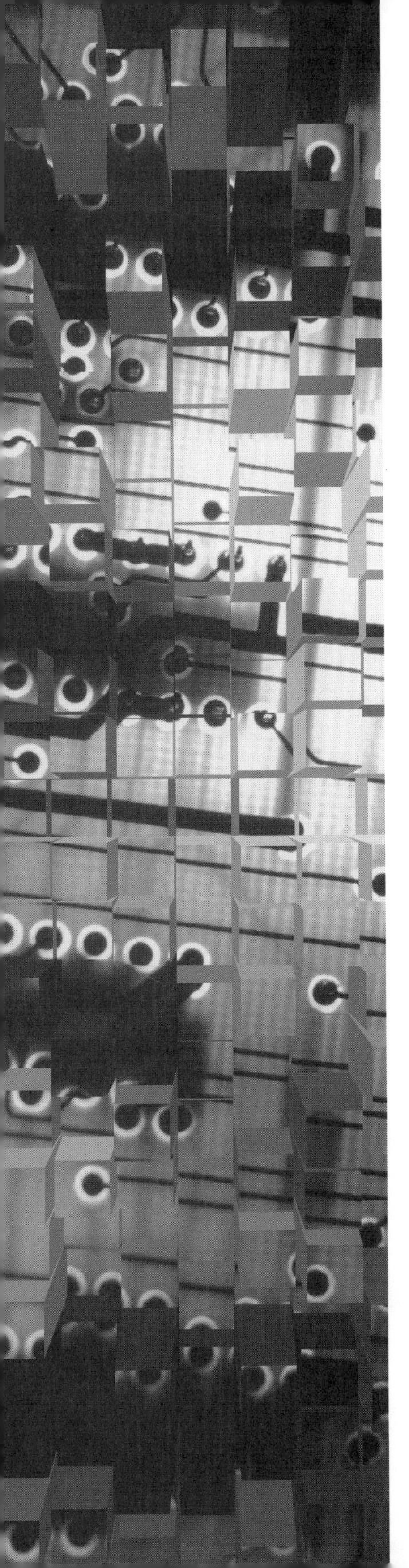

Foreword

The worlds of content, commerce, and advertising on the Internet are colliding and intersecting, unlocking tremendous new opportunities and revenue streams. Until recently, it was difficult to exploit the possibilities afforded by this change due to a lack of standard formats, protocols, and tools. Now, tools are available that accelerate the development and encourage the convergence of the previously discrete worlds of content, commerce, and advertising.

Foundations for this robust, "next-generation Internet" are just now being laid: cross-platform programming languages such as Java; markup and stylesheet languages such as XML and XSL; protocols such as ICE for syndicating content and WDDX for exchanging dynamic applications; metadata languages for publishers such as the PRISM initiative. Each initiative is a subset of the larger dynamic taking place—the new infrastructure is enabling deep, rich, dynamic content, advertising, and commerce relationships between online partners.

Ultimately, Web sites are going to be expertly crafted "user experiences." Users will exchange their valuable time to enter into experiences that have timely, relevant, and compelling information. The best sites create these experiential worlds by aggregating content, commerce, and advertising from a host of creative and unique sources. For example, almost all of the data on Yahoo!'s site is obtained from other source sites. Yet Yahoo! maintains strong customer loyalty and interaction by weaving disparate content from multiple syndicators into uniform, personalized, and contextual end-user experiences.

The intersections of these colliding worlds are unlocking new business models, products, and services. Standard protocols will be used to syndicate the *buy* button—a merchandiser can create digital content and embed business logic into it; for example, control of brand loyalty, digital property rights, or financial audit trails for royalty payments. This content can be syndicated to third-party sites without any loss of control. Business rules can be embedded into "advertorials" that combine commerce, content, and advertising into one digital asset. Essentially, the merchandizing of content allows publishers to turn content into a product that can be cataloged, inventoried, and sold.

The timing of *Killer Content* is perfect. As killer content evolves, and what constitutes "fair exchange" for a user's time, loyal patronage, and purchasing power, it is undergoing rapid and profound change. Content providers that understand and leverage this fundamental shift can merchandize and syndicate their digital assets to take advantage of new pricing, advertising, and distribution models. These new models unlock powerful and diversified revenue streams on the Internet that were previously unimaginable.

David Mathison
Chairman, CEO, and Co-Founder
Kinecta Corporation
San Francisco

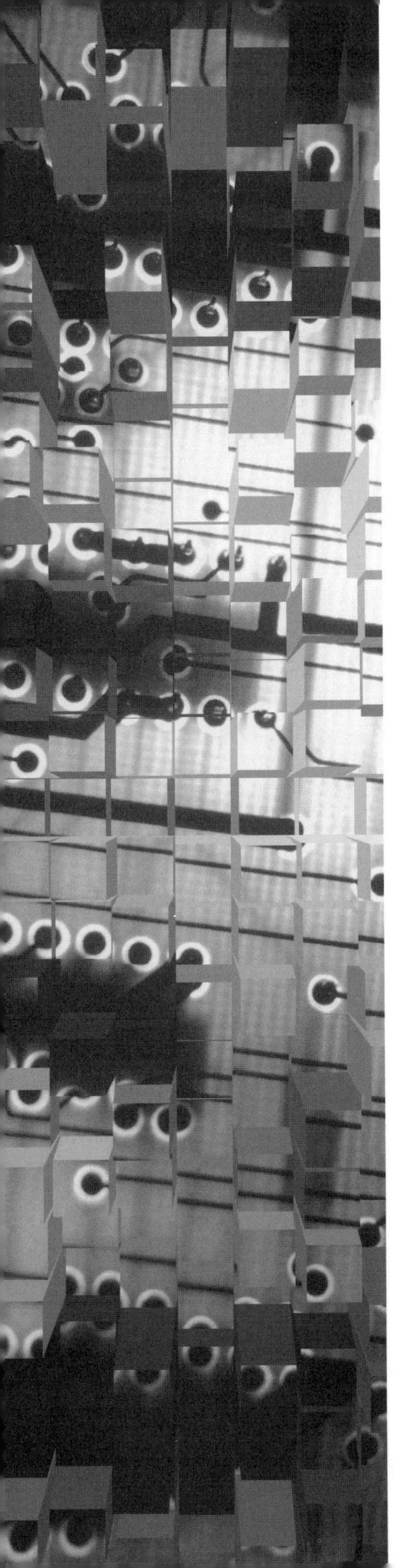

Preface

It occurred to me one day, as I was reading the morning news online, that I had turned into quite the creature of habit on the Internet. This surprised me. I always thought that I would be the first one to try something new on the Web. I am what analysts call "a technology optimist," which is just another way of saying that I'm prone to trying out the latest cool digital toy or service.

My interest started back in the early days of e-commerce when I worked on the first version of their Microsoft's Site Server Commerce Edition. I have used the Internet for at least one hour a day since 1995. I hand over my credit card to online merchants on a regular basis for books, airline tickets, pet supplies, and far too many plants for my city garden. I have come close to but have not quite arrived at using a digital subscriber line (DSL) for high-speed Internet access from home. And, when our 1990 Honda Civic hatchback finally sputters to a halt, we will probably buy our next car online. Yet despite my familiarity with the Internet, I do not idly browse for new content or sites. No, I actively *use* the Web in what I realized is a structured manner.

I use it for two purposes. My first goal is to increase my free time by truncating the effort involved in household errands such as buying food for the pets. My second objective is to maximize my ability to understand the world around me, both in personal and professional terms. Underpinning both reasons is the assumption that I can find what I need quickly and easily through a Web site. If I can't, I move on to another site that gives me what I want in price, selection, information, or service. No ambiguity about it. I don't have time to spend on out-of-date information, slow online services, or byzantine page flows through a Web site. So I don't return—or at least I have never returned—to a Web site that doesn't immediately provide me with enough value for my time and money.

I am part of a Web community of experienced Net users. We want more than just a good price and three-day delivery. We're looking for a user-centric experience. When I visit a Web site, I'm subconsciously thinking, "In exchange for my time and potentially my money, I want content and services that are relevant to *me*." We are part of a growing trend among Net users who look for something beyond price point or brand in our Web-browsing experiences.

This book is a primer for Internet content business models that address Net users' demands for additional value and services. In Part One: Concepts, I explore the definition and real-world implementations of value exchange, premium content and services, and the emergent selling and payment models on the Web. As new business models develop, the line between content and commerce blurs. Merchandizing content allows publishers—companies and organizations that have, or plan to set up, their own Web sites—to explore new revenue streams generated by their core competencies.

Part Two: Strategies focuses on what Web publishers can do to enhance and monetize value exchange. These strategies help publishers evaluate the practical steps for implementing added services for their own Web sites. Taken as a whole, this discussion looks beyond the world of retail transactions to an Internet economy where a Web publisher's "product" can be a book, an auction environment, an expert opinion, or an Internet radio broadcast.

This book has two audiences. On one hand, I provide content and commerce Web site publishers with concepts and strategies to help them improve business-critical conversion rates (i.e., the rate at which a casual visitor becomes a loyal visitor, a loyal visitor turns into a first-time buyer, or a first-time buyer becomes a repeat customer). There's always talk about the next Internet "killer app"—a concept or product that has a revolutionary impact on the online world. The first commercial browser was a killer app, as was e-mail. Improved and monetized content—a site-specific concept that galvanizes the growth of revenue and traffic—constitutes a killer app in its own right. "Killer content" satisfies the goals and objectives of visitors in exchange for loyalty or buying power. This value exchange between publisher and visitor drives conversion rates and viable business models for the Web. Without the ability to enable and monetize successful value exchange, Web sites bleed money without gaining the Net user base on which to build advertising and commerce revenue.

This book is also for Net users who are interested in learning about new types of Web content and commerce. As premium content and services become more available, Net users leverage the different types of value exchange to make the Internet more valuable at work and at play. An enhanced awareness of high-quality content and services for users might galvanize a switch to a Web site that offers personalized profiles for quicker purchasing, a bid in an online auction, or a first purchase of premium content.

Tim Berners-Lee, the inventor of the World Wide Web, says in his essay "The World Wide Web: A Very Short Personal History" that "the dream behind the Web . . . is dependent on the Web being so generally used that it becomes a realistic mirror [or in fact the primary embodiment] of the ways in which we work and play and socialize" (7 May 1998, *www.w3.org/People/Berners-Lee/ShortHistory.html*). For Net users, the value lies in "milking" Web sites for maximum benefit, whether it's a live broadcast feed of the ball game or insight into the hidden charm of an undervalued stock. This book illustrates the ways that the Web is evolving toward that "realistic mirror" that Berners-Lee prophesies.

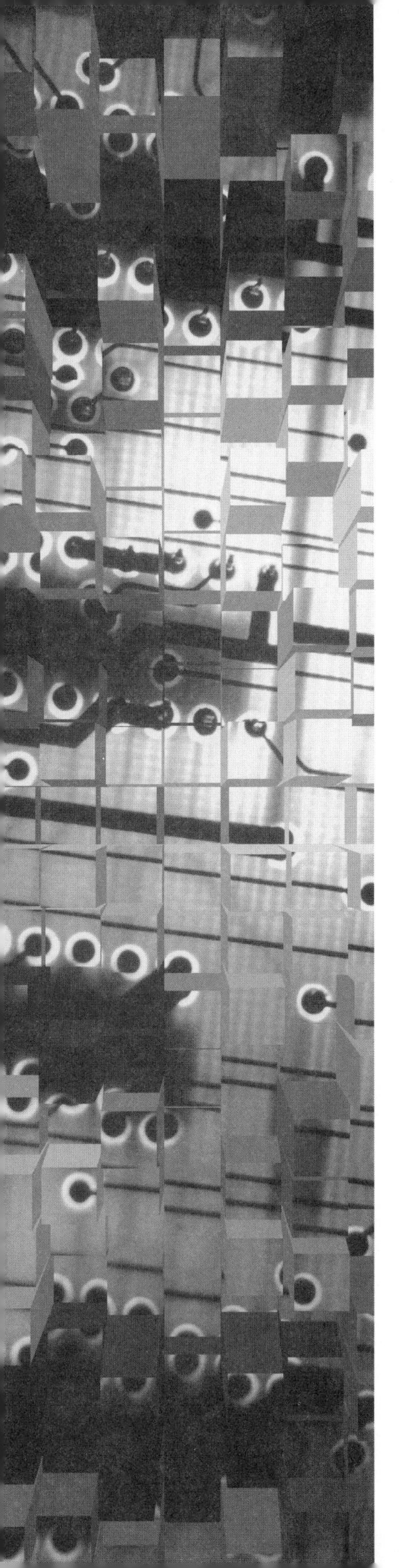

Acknowledgments

Now that the book is in readers' hands, I can safely say that writing a book was a lot more fun that I expected. It was also considerably more work. If I hadn't had help along the way, *Killer Content* would be a far different read. I want to thank my family, former colleagues, coworkers, and new friends who have added their industry experience and insight to the evolution of this book.

First, I'd like to commend my family and friends for their patience and good nature in the past months. My best friend and husband, Mark, gets the gold medal for dealing with writer's block and providing constructive advice and an encouraging voice as deadlines approached (again and again). There's no small element of courage required for that task. My parents, Ong Ba Ngoai, and sister have responded wonderfully to the time demands of writing a book. I appreciate your unfaltering support over the years. (My mother has been waiting a long time for the book.) This book is dedicated to my entire family for their love and support during the months of writing.

Killer Content has also benefited immeasurably from the insight of new friends and colleagues in the Internet space. I'd like to call out the ones who played the most significant role in shaping this book, although many others contributed to its evolution. First, many thanks to David Mathison for taking time out of his busy schedule in building a successful start-up to ponder some of my ideas and write the foreword for this book. Kirk Loevner, Ron Faith, Jeffrey Bell, Jeremy Allaire, and Erik Josowitz provided excellent advice throughout the research and writing. I appreciate the knowledge and experience you provided in our discussions.

My reviewers (Robert W. Husted, Bill Radcliffe, Ronald A. Radice, Robin Rowe, Richard Careaga, Michael R. Blaha, J. J. Kuslich, Gary Erickson, Tom Valesky, Sanjiv Gossain, and Steve Pellegrin) have done much to help transform those first early drafts into the book's current form. Thanks for your thorough reviews and cogent feedback. My editor Mary O'Brien and editorial assistant Mariann Kourafas shepherded the manuscript through its various phases with expertise and grace. Their professionalism demonstrates the truth behind Addison-Wesley's reputation as an excellent publisher.

My development in two areas have shaped this book and my career as a product and program manager in the Internet field: writing and distributed software technology. On the writing front, I want to thank Fanny Howe and Mel Freilicher for their steadfast support and encouragement when I was first getting my writing legs. Within the software industry the friendship and mentorship of Seth Manheim, Robert Barnes, Tracey Trewin, Hugh Teegan, and Ron Faith have contributed both directly and indirectly to my ability to write this book.

Again, my thanks to everyone who helped make this book possible.

PART ONE

Concepts

Part One describes the elements of killer content and explores the theory behind established and emerging business models for value exchange. It discusses the business logic behind enhanced value exchange and the role of relevance in the browsing experience. As the line blurs between content and commerce sites, Web publishers use new strategies, such as affiliate programs and personalization, to foster loyalty to sites.

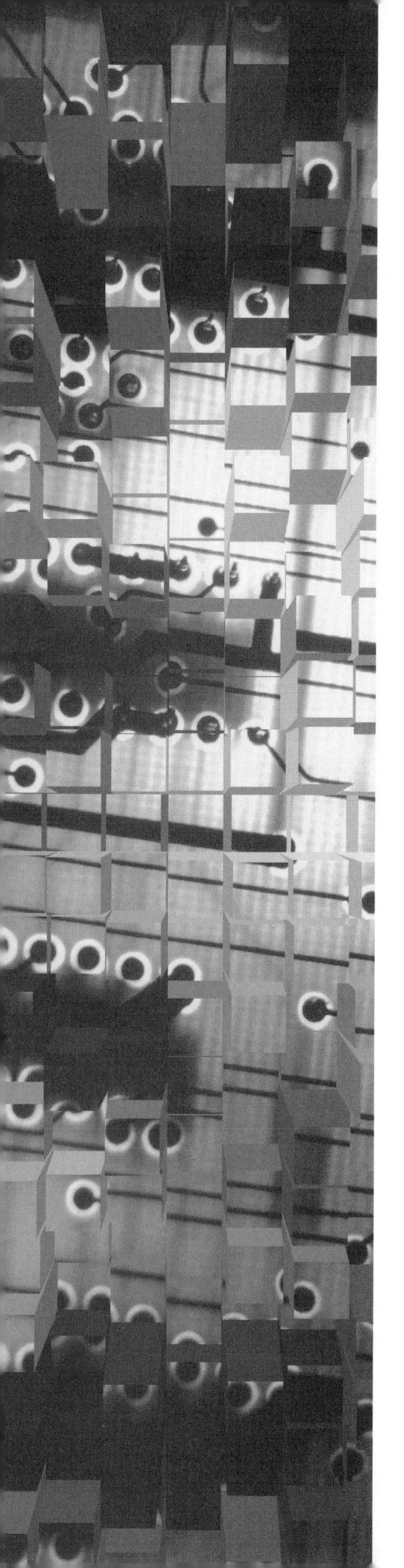

CHAPTER ONE

Content Value Exchange

Whether you're a content publisher, e-commerce merchant, or publisher of a promotional Web site, your success on the increasingly crowded Internet depends on converting a casual site visit to something that feeds your business model.

If your revenue hinges on advertising, as many content sites do, you need high volumes of first-time and return Net visitors. If you're a Net vendor, you care about maximizing the number of transactions that flow through the Web site. If you host a promotional Web site to generate leads for your business, you want your Web site to act as a calling card for your credibility, brand, and reputation. Whether you combine advertising and commerce, or not, you have a real dependency on providing a high-quality user experience to your visitors. Turning "fly-bys" (one-time visitors) into repeat users and online window shoppers into paying customers are critical in light of significant Web site operating costs and razor-thin margins for inventory. These "conversion" problems can seem maddeningly difficult to solve and constitute the largest obstacle to revenue for Net businesses. It's not cheap building a Net user base on the Internet. It's essential for publishers to retain new customers through existing services provided by the site.

The answer to acquiring and keeping Net users does not necessarily depend on the most expensive e-commerce system or hip graphics. The real solution is buried in the relationship, or value exchange, formed between the Web site and visitor around killer content.

Concepts and Strategies

At its most basic level, service enhancement improves the value exchange between publisher and visitor. The end result is added incentive for a Net user to return to the site or to make the first (or second) purchase. Service enhancement and value exchange are not radically new concepts. Both concepts derive from the basic premise of customer service. What *is* new is the idea that many of the problems that Web sites face today with elusive visitors and customers can be solved with quality content and services rather than complex technologies. In the Internet economy, a "killer app" is a concept or application that enacts radical change in the online world. E-mail and Web browsers are both killer apps, since they fundamentally changed the way people interact and use the Internet. On a site level, any Web publisher can build his or her own killer app by enhancing the quality of the user experience through *content and services.* These enhancements impact Web site design, infrastructure, and business models. Providing killer content on top of a site's core competency shifts the model from a single dimension, such as e-commerce, to a multilayered relationship between that is able to withstand the Internet price wars.

The nature of a value exchange around content depends on the goals and objectives of both the Web publisher and visitor. The first step in designing value exchange is to understand the elements that constitute the relationship and to gauge the relationship's impact on conversion ratios. Value exchange looks very different on general news Web sites, such as CNN Online (*www.cnn.com*), than on a purely promotional site.

Drink manufacturer Snapple uses its Web site (*www.snapple.com*) to promote the advertising campaign themes run in print media such as magazines. Web site visitors expect to be entertained or win free merchandise at a promotional site like Snapple's. In contrast, visitors to CNN Online use the news site's content to learn more about current events' headlines and breaking news. Different online value exchanges demand different business models. Sites that blend content and e-commerce demand new models that break down the previously clear line of distinction between e-commerce vendors and content publishers. Value exchange on the Web is no longer limited to the traditional e-commerce exchange of credit card payment for a retail item. There are new types of payment, delivery, and product to add to the mix. Just as multiple layers of service enhance a site's core competency, business models benefit from diversity. Advertising, e-commerce retailing, syndication, and pay-per-access premium content represent just a few of the new hybrid business models already at play on the Web. As payment methods and the notion of what constitutes a product become increasingly flexible, the world of commerce extends beyond retail goods and into content. And at the heart of killer content lies the notion of value exchange.

The Need for Value

Internet sites look perilously similar when searched for content and products. A Net user looking for the day's headlines can travel between half a dozen news-oriented Web sites before realizing that each site provides very similar content—the same stories, the same static presentment of information. Online stores offer the same inventory at similar prices with little variation to differentiate one site from another. The brand and location of the site may differ, but the actual commodity does not vary significantly. The core difference in Web sites lies in the *additional* value that the Web publisher does or does not provide as part of the purchase or browsing experience. That value can take a variety of forms, ranging from deeply discounted prices to in-depth commentary on mainstream news headlines.

For a minority of Net buyers (users who purchase online), the primary goal is simply to obtain the desired commodity as quickly as possible (see Figure 1.1). These users turn to the Web for its efficiency in collecting orders

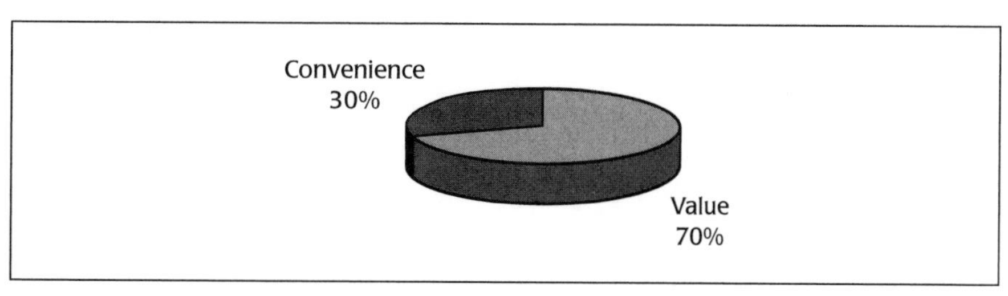

Figure 1.1 Primary Motivator for Online Purchases (estimate for year 2002). Copyright © 2000, Forrester Research, Inc. Used with permission.

and delivering content. Convenience-oriented users care more about the speed at which the value exchange occurs rather than the individual elements of the exchange (such as price, customer service, or relevance of experience). Convenience drives relevance for these users in the online experience.

But according to a 1999 Forrester Research report ("Are Net Shoppers Loyal"), the majority of Net buyers (an estimated 70% by 2002) focus on value. All Net users search the Web for content and products that provide the highest degree of relevance for their purchasing and browsing criteria. A Net buyer concerned with the lowest price for a given product has cost as his or her most relevant criterion. The Net buyer can also have other less relevant but still valued criteria, such as overnight delivery or a clear channel for customer communication. These factors and degrees of relevance differ by content, product, and Net user. A Net buyer for subscription financial information highly values the credibility of the source of content, which the buyer relies on to make financial decisions. In contrast, a Net browser of a popular extreme sports Web site cares about the innovation in presentation of content—for example, rich media displays such as broadcasts of snowboard tournaments or streaming video clips of resort sites.

Regardless of the criteria, a lack of relevance in user experience discourages a return visit. From the Web publisher's perspective, producing a Web site without any distinguishing characteristics reduces the ability to create sustained traffic and commerce from loyal Net users. A Web site that does not immediately resonate with a Net user will be quickly abandoned in search of more interesting content and services. The Internet is getting crowded with Web sites that offer very similar experiences. Take a look at the Web and you'll see many sites that present information and sell products in a manner that can be and are easily reproduced by the online store "down the block" (or at least, only an online shopping network away). The publishers who stand to reap the most benefits from e-commerce are the ones who add relevance to their Web site through value exchange. Unfortunately, there are only a limited number of Web sites that offer real value exchange to Net users today.

The promise of trillion-dollar revenue from e-commerce has driven many existing or would-be publishers to lose sight of the real promise of the Internet—a mutually profitable quid pro quo between Web site and Net user. American poet Robert Frost once wrote about how man can turn the noise of the wind into song. "Man came to tell it what was wrong," wrote Frost. "It hadn't found the place to blow; It blew too hard—the aim was song."[1] The Internet right now is much like a noisy wind, blowing information, commerce, and content indiscriminately at Net users without concern for preference or personalization or other user-driven value factors. Without value exchange, publishers drastically undercut their opportunity to convert casual or first-time Web users to frequent browsers and buyers.

Value exchange adds relevance to the Net by carving out distinct relationships between Net users and Web site content and services. *Value*, according to the *Merriam-Webster Dictionary*, can be defined as "a fair return or equivalent in money, goods, or services for something exchanged." As the Internet evolves, Net users shape the evolution of the Internet with their views on what constitutes a "fair return" for their time and patronage of a Web site. Net users demand a service relationship from Web publishers. Publishers are starting to put in place the tools—such as personalization and collaborative filtering—that support that value exchange.

Web sites that make the user experience relevant and useful for Net visitors can convert casual Web users into an integral part of the Internet economy. Whatever your aim for your Web site—whether it's promotion, content, commerce, or entertainment—adding elements for value exchange builds a base of loyal Net browsers and buyers.

The Dilution of Loyalty

The lack of Net user value on Web sites can be attributed to some degree to the incredibly fast evolution of the Internet as a commercial exchange. Back in the 1960s, the Internet was still a closed US Department of Defense experiment in building computer networks. Only defense contractors and academics had access to the Advanced Research Projects Agency Network (ARPANET). Content and services in those days consisted of primarily text files, newsgroups, and binary file exchange. ARPANET was about information exchange, not user experience.

---

[1]From "The Aim Was Song," in *The Poetry of Robert Frost*, edited by Edward Connery Lathem. Copyright 1915 by Robert Frost. Copyright © 1979 by Henry Holt and Company, LLC. Reprinted by permission of Henry Holt & Co., LLC.

Terminology

Content: Text, images, audio, and video that compose your Web site.

Conversion: Transformation of a casual visitor into a frequent visitor, a visitor into a first-time buyer, or a first-time buyer into a repeat customer.

Digital goods: Content or software distributed through the Internet. Digital goods can mean the latest upgrade of a software program downloaded from the Web or the MP3 file for The Chemical Brothers' *Block Rockin' Beats* song.

Hard goods: Physical commodities sold over the Internet and distributed via supply–distribution chains.

Net browser: Net user who travels to one or more destinations on the Internet.

Net buyer: Net user who makes a purchase through one or more Web sites.

Net user: Someone who uses the Internet for Web browsing or e-mail.

Online: Actions that occur on the Internet. Conversely, the term *offline* describes items in the physical world.

Online services: Services offered to help Net users perform an action, such as holding payment in escrow for a buyer and seller in an online auction or comparing prices automatically on the Web.

Web publishers: Anyone involved with developing and producing a Web site, whether a senior manager, a Web developer, or a copywriter.

Things changed radically with two major events: the government was pressured into opening its private network to the public, and students and faculty at the National Center for Supercomputing Applications (NCSA) started writing Web browsers. Some NCSA students and faculty, including one student named Marc Andreessen, wrote the first Net user-accessible Web browser called Mosaic,[2] which was released for Net user platforms in September 1993.

[2] Other browsers, such as ViolaWWW, Erwise, and Midas, preceded Mosaic. However, Mosaic was the first browser that was easy to install and use, got online images to work, and provided customer support.

Andreessen and some other Mosaic programmers left to start a little company named Netscape, and the Internet race was on. According to the Online Computer Library Center (OCLC), Office of Research, by 1999, the Web had grown to approximately 3.6 million unique sites, at a rate of at least 100% growth per year. A full 61% of those sites offer publicly accessible content,[3] totaling nearly 300 million Web pages in all.

The number of destinations does not obscure a single fact: very little on the Web, in terms of content, services, or product, is truly unique. With inventory, anyone can build an online bookstore. With a network infrastructure plugging into airline reservation systems (and plenty of capital), anyone can build a self-service travel-reservation-booking site. And certainly anyone with an opinion can set up his or her own editorial/commentary online magazine (or *webzine*). So if nothing is unique (or alternatively, any business can be replicated) on the Internet, what can a publisher offer to earn a Net browser's loyalty and buying power? The answer lies in the enhanced services layered on top of the Web site's core competency.

Net user control on the Internet translates to customers who are perfectly willing to depart at any time, who have a low frustration threshold, and who demand a high level of service in exchange for loyalty. "It is becoming increasingly clear that the Net user decision-making process is different," said Professor Hoffman, the codirector of the Electronic Commerce Research Center at Vanderbilt University in a 1999 interview with *The New York Times*.[4] "And when we look at firms that are making mistakes or not quite getting it right, in part it can be attributed to their not understanding that cyberspace is not a simulation of the real world" (Goode, 1999). By paying attention to value exchange, a publisher provides Net users with real reasons to visit and return consistently to a destination.

Any Web publisher has the ability to establish a relationship with a Net user around content. Content Web sites that provide online information and services encourage visitor loyalty and drive revenue through the innovation, relevance, credibility, and timeliness of content. E-commerce Web sites that

---

[3]These statistics do not include "provisional" (unfinished or trivial content) Web sites. (Information from Jan. 5, 2000, statistics regarding the growth of the Internet on the Deep Canyon Web site—*http://www.deepcanyon.com/b/numbers/nn10041999.htm.*)

[4]Erica Goode, "The Online Consumer? Tough, Impatient and Gone in a Blink." *New York Times,* 22 Sept. 1999.

Five Value Factors for Net Users

Credibility: Authority of the source of the content, product, and/or service.

Innovation: Uniqueness of content, product, and/or service.

Relevance: Impact on a decision, goal, or lifestyle.

Timeliness: Immediacy for time-sensitive operations.

Utility: Usefulness in performing daily activities.

collect orders for physical goods establish loyalty by offering shopping experiences that are intuitive and offer credibility to the user.

Early Examples of Value Exchange

The first types of commercial value exchange on the Internet occurred between retail merchants and Net buyers. Amazon.com and Cisco represent the "best of the breed" for early adopters of value exchange in their Web business models. Both realized in the mid-1990s that value exchange was the key to commerce. In fact, a shared concern for value exchange and an early start date in the e-commerce race are some of the few things that Amazon.com and Cisco have in common. Let's first take a look at how value exchange has evolved at Amazon.

Net User Value Exchange

Since Amazon.com first launched in 1996, the Net retailer has prioritized company growth (including service enhancements) above profits. Jeff Bezos founded the company in 1994 on the conviction that the Internet was perfect for selling books because book distributors tracked inventory through well-maintained electronic lists. Amazon.com appeared as an Internet-only bookseller on the Web in July 1995. Keep in mind that in 1995 not much was happening in the field of e-commerce, especially in the business-to-consumer space.[5] The browser war between Microsoft and Netscape had yet to be

[5] *Business-to-consumer* (or B to C) refers to merchants selling directly to end users. *Business-to-business* (or B to B) commerce involves merchants selling goods and services to other merchants.

launched. Banks were beginning to hype the promise of secure encryption technology (SET) to protect Net user credit card information. Amazon came into the e-commerce arena with a single mission: to sell as many books as possible.

Amazon adheres to a central distribution model for goods sold through the Web site and actively manages the scale aspect of the site so that costs do not rise proportionately to sales. Because the company collects payment before ordering and distributing goods, it has a strong cash-flow system. Customers pay for their orders well before the company pays suppliers. Since 1995, Amazon has moved beyond books into other branches of Net user retail, such as electronics, music, and video. With every move into new Net user territory, Amazon has carried the banner of superior Net user service. With Net user-friendly techniques like the convenient One-Click site shopping cart (patent awarded in October 1999), lenient refunds, and frequent upgrades for shipping, Amazon has built what is perhaps the most respected Net user base in e-commerce today. More than 10 million Net users shop at Amazon.com. That's in excess of 10 million people who have provided the company with their credit card information as well as some subset of personal preference and purchasing patterns. While not rich in profits, Amazon.com is incredibly wealthy in customer loyalty and consistency.[6] Without their early start in e-commerce, Amazon.com could never have obtained such a strong customer following by the time that most online booksellers, such as Barnes&Noble, were just launching their Web sites.

Amazon.com uses its well-nurtured relationship with customers as the basis of a Web community that focuses on service. That emphasis on service permeates the Web business, from responsive customer support to an efficient, easily navigated Web site. From the first two months' profit of $20,000 a week back in 1995, Amazon.com continues to plow profits back into enhancing the usability of the Web site.

Amazon's emphasis on customer service promotes innovation in streamlining the purchase process. In fact, Amazon.com came up with many of the Net user services now seen across the Web, such as one-click shopping, customer reviews, and e-mail order verification. To help target recommendations

---

[6]Actually, Amazon.com had yet to turn a profit by the end of the second quarter of 1999. Amazon's sales trajectory is aggressive, though: sales increased 283% from the fourth quarter of 1997 to the fourth quarter of 1998. *Source:* Hoover's Online.

for customers, Amazon uses *collaborative filtering*, which analyzes purchase patterns and preferences, matches a Net user up with other Net users with similar buying habits, and makes recommendations based on the match (see Figure 1.2).

Jeff Bezos recognized early that relevance in retailing options and services marked the key market advantage over new competitors. As a result, Bezos spent much of the late 1990s acquiring online services, such as Junglee, a comparison-shopping service, and PlanetAll, an address book, calendar, and reminder service. Bezos didn't stop at just services. Other full or partial acquisitions include drugstore.com, HomeGrocer.com, Pets.com, Gear.com, Web-use tracking firm Alexa Internet, auction site LiveBid.com, rare book and music marketplace exchange.com, and e-commerce systems firm Accept.com. Amazon's spending spree has delayed the company's profitability but deep-

Figure 1.2 Personalized Recommendation from Amazon.com.

Source: Copyright © 2000, Amazon.com. All rights reserved. Used with permission. Amazon.com is a registered trademark or trademark of Amazon.com, Inc. in the United States and/or other countries.

ened their market advantage over retailers who focus more on inventory and price than services.

Amazon has also pioneered marketing tactics such as the affiliate program. Amazon allows both merchants and consumers to sign up as Amazon affiliates. Once they are affiliate partners, Web site publishers post links to Amazon products on their personal or commercial Web sites. Amazon then pays the affiliate partner a percentage of any sales resulting from a consumer clicking through on the link. Affiliate links' sales constitute a growing percentage of Amazon's total revenue. By the middle of 1998, Amazon.com's market capitalization equaled the combined values of profitable offline stores Barnes&Noble and Borders (whose combined sales exceeded Amazon.com's).

The wide array of services has turned Amazon into the number one site in building customer loyalty, according to an October 1999 Peppers and Rogers study. The study examined factors in the value exchange such as how well a Web site identified its customers and prospects; how well it determined need and value; how effectively customer interaction was conducted; and how well it customized the user experience. Amazon scored higher in all categories than other e-commerce sites, spanning the retail online market (Peppers and Rogers, 1999).

Amazon.com Value Exchange

For Net Users: Personalized recommendations, competitive prices, lenient return policy, responsive customer service, recommendation boards, wide retail selection, special buyer rewards for frequent customers, and cross-site promotions with partner stores.

For Amazon.com: High browsing traffic, popular affiliate program, high volume of sales, and extensive database of personal information on customers.

Value Factors:

Credibility (high brand recognition)
Price (low cost for books)
Relevance (wide variety of services and products)
Utility (easily navigated site, search functionality, and streamlined purchase process)

Business-to-Business Value Exchange

Enhanced services in business-to-business exchange have the same goal—to ensure customer satisfaction. This often translates to efficiency in order placement, tracking, and fulfillment.

Cisco Systems, one of the first providers of business-to-business value exchange on the Web, provides a highly automated and efficient system for order collection and supply-distribution through its Web site. While Net users may not have heard the name Cisco, they have probably used a Cisco product while browsing the Web. The company, which is based in San Jose, California, is the leading supplier of computer network products and a rising player in telecommunications networking. Cisco sells the tools that make the Internet (and company intranets) run, such as routers, switches, software, and other components of network infrastructure.

Since Cisco sells products to other businesses, it focuses on the elements of service that resonate most significantly with other companies: reliable order fulfillment and efficient customer service. The Cisco Web site provides a complete set of services for the small, mid-size, and large companies that use Cisco products (see Figure 1.3). John Chambers, Cisco's CEO, intends to make Cisco the model of an efficient and customer service–oriented Net company. According to Chambers, almost three-fourths of all orders and customer support are transacted via the Web (Daly, 1999).

The Web site also serves as a sales channel for the company (Cisco currently sells a full 78% of its products online). A customer can place an order, monitor the status of the order, and receive quick customer service. Automating the order collection and fulfillment process streamlines the delivery of the product to the customer, reducing error and eliminating unnecessary steps along the way. By integrating members of Cisco's extensive supply distribution chain with the order tracking process, Cisco has both shortened the delivery time for network components as well as ensuring increased visibility of each step in the process. After the quick delivery of their order, customers also get immediate answers to questions from Cisco's automated customer service bureau (seven out of ten questions are able to be answered via the automated service).

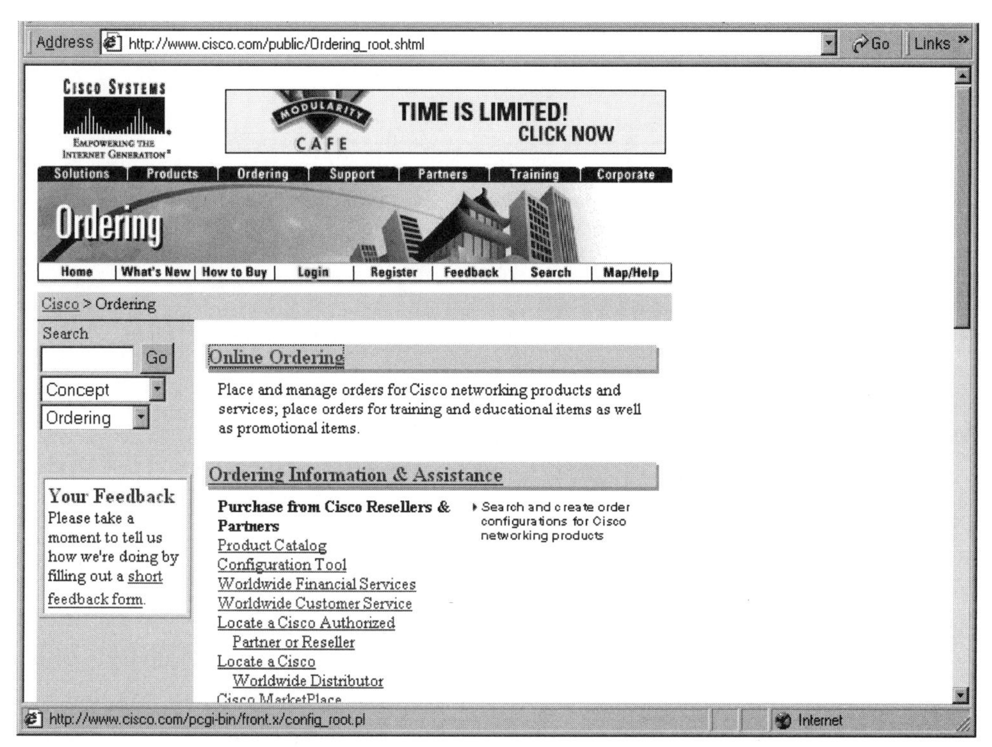

Figure 1.3 Ordering, Tracking, and Customer Service on Cisco's Web Site.
Source: These materials have been reproduced by Addison-Wesley Professional with the permission of Cisco Systems, Inc. Copyright © 2000, Cisco Systems, Inc. All rights reserved.

Common Success Factors

While Cisco and Amazon have different types of customers and products, both companies share a common belief in enhanced services as the key to continued success. Look at the other commerce leaders on the Internet, and, for the most part, you see companies that take their core competencies seriously by providing low prices and wide choices in exchange for their customers' business. But their edge comes from enhanced value exchange—services such as superior customer service, flexible shipping, effective distribution, communities—and usability enhancements such as personalized recommendations.

The Web sites that offer *relevant* and *useful* content and services offer a more compelling value exchange than destinations with similar core compe-

Cisco Value Exchange

For Net Users: Self-service for order placement and tracking, automated customer support, in-depth information about the product line, quick response to the customers' requests, and reduced human error in ordering.

For Cisco: More efficient order fulfillment, shorter time to market, higher sales, and lower cost center for customer support.

Value Factors:

Relevance (availability of an extensive inventory)
Utility (order submission, fulfillment, and tracking through automated systems)

tencies but no commitment to relevancy and usefulness. This distinction between core competencies (like low-price product inventories and current news headlines) and added services (such as personalization or collaborative filtering) becomes more important as more potential buyers migrate to the Internet.

It's no secret that Internet usage will continue to increase dramatically in upcoming years. New York–based Internet research and advisory firm Jupiter Communications, and NFO, predict that about half (56%) of the US population are likely to be online by 2003. That translates to potentially 157 million people online in the United States alone by 2003 (Jupiter, 1999).

Maximizing the opportunity requires an understanding of the nuances involved in value exchange. Evan I. Schwartz in his latest excellent study of the evolution of the Web (*Digital Darwinism*) notes that Web site vendors cannot profit without "identifying a specific set of issues that . . . customers face, then developing a set of interactive services that address these problems" (Schwartz, 1999). These "interactive services" turn Net browsers into Net buyers and encourage loyalty

Net Browsers and Net Buyers

There are two types of Net user: Net browsers and Net buyers. Net browsers use the Internet to read news, research, e-mail, and perform other activities that do not require a payment method. Net buyers purchase goods, which range from actual commodities, such as electronics, to premium content through paid access, such as subscriptions. There are far more Net browsers than there are Net buyers—almost three times as many (see Figure 1.4).

Despite the ratio of buyers to browsers, there is still a belief that the only type of value exchange on the Internet involves the collection and fulfillment of retail orders in exchange for credit card payment. Many novice Web publishers believe that the only way to exchange value with Net users is to offer an inventory of desirable products, an order collection mechanism, and a corresponding supply/distribution chain.

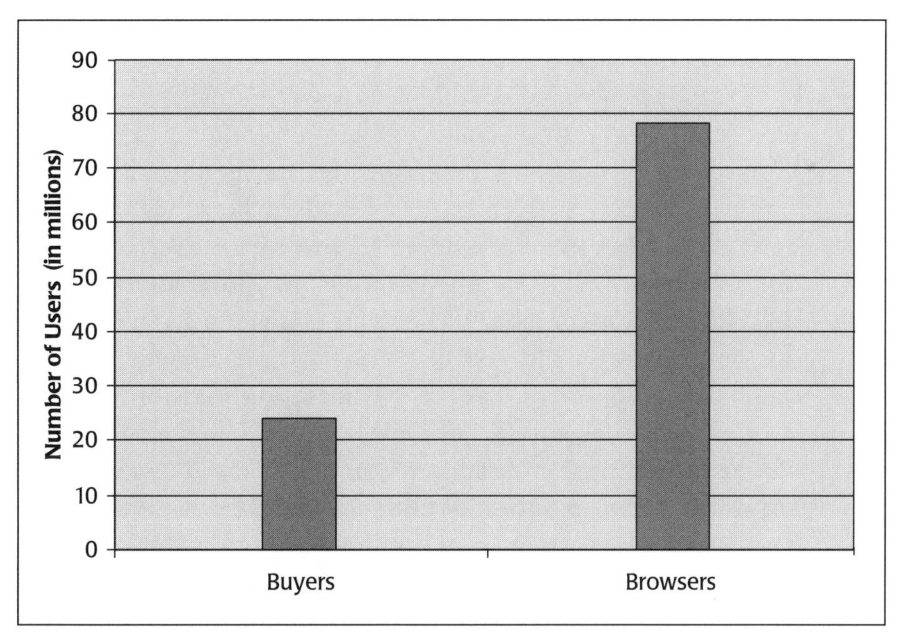

Figure 1.4 Ratio of Online Browsers to Online Buyers.
Source: From "Attitudes, Behaviors and Demographics of the Online User." The Jupiter/ NFO Consumer Survey, Vol. 2 (p. 25), August 1999. Copyright © 1999 Jupiter Communications. Used with permission.

Net Shoppers' Key Demographics

Average income: $60,000

Average age: 30.7

Source: Information from Jupiter analysis in "Attitudes, Behaviors and Demographics of the Online User." The Jupiter/NFO Consumer Survey, Vol. 2 (Figure III.5, p. 29), August 1999. Copyright © 1999 Jupiter Communications and NFO Interactive. Used with permission.

In reality, a Web publisher must establish a value exchange with Net browsers in order to motivate Net purchases. Let's take a look at how a Net browser becomes a Net buyer. (We'll review a site that offers value exchange through free as well as premium content to give a sense of the different elements in the value exchange.)

On an Internet crowded with sites, content is king. Consider the sheer volume of sites currently available: roughly 60% of *all* medium- to large-size businesses worldwide will establish a Web presence by the end of 1999, according to statistics site eMarketer. Hundreds of thousands of Web sites jockey every day for Net user bandwidth and attention across Internet and traditional channels. Television commercials and magazine advertisements ask Net users to check the corporate Web site for more information about a promotion or a contest. On the radio dial, Web sites bring us news updates. Baseball stadium placards now include brand names preceded with "www."

Despite the ubiquity of a "dot com," or Web, presence, successful value exchange on the Internet depends on providing something that satisfies Net browser or buyer objectives for using the Web. Merchants and content publishers alike attract first time visitors through brand credibility fostered by advertisements or public relations. These Net vendors retain visitors through a combination of inventory, price, and service. The shift from Web browser to buyer (or one-time visitor to repeat user) occurs when the publisher offers something of strong relevance to the Net user. For users who crave a good deal, a low price might promote the transition.

Overnight delivery might attract other Net users who need the product immediately as a gift or for household use. If the Net buyer ends up returning to the Web merchant for multiple purchases, the publisher has obtained sustainable revenue from the exchange. To retain frequent purchasers, publishers maintain a high level of service (delivery, returns, and exchanges) as well as providing improved user experiences (for example, purchase rewards and personalization).

Every step of the evolution of a Net buyer (see Figure 1.5) involves the provision of some form of value to the Net user. If something at the site catches the Net visitor's interest, the visitor might stop by again during another browsing session. Building loyalty depends on the quality and relevance of the content. Offering the same headlines as a half dozen other Web sites may, at best, generate casual interest from a Net browser. In contrast, offering personalized headlines and commentary encourages repeat visits and customer loyalty.

Loyal Net browsers provide value through sustained traffic for advertising-dependent sites and also increase the pool of potential buyers for premium content or retail products. The leap to Net buyer occurs when the content, service, or product strikes the right note of relevance for the Net user. That degree of relevance can be in cost (*the lowest price on the Web*) or in uniqueness (*the only place on the Web*) or in customer service (*the quickest and most reliable site on the Web*).

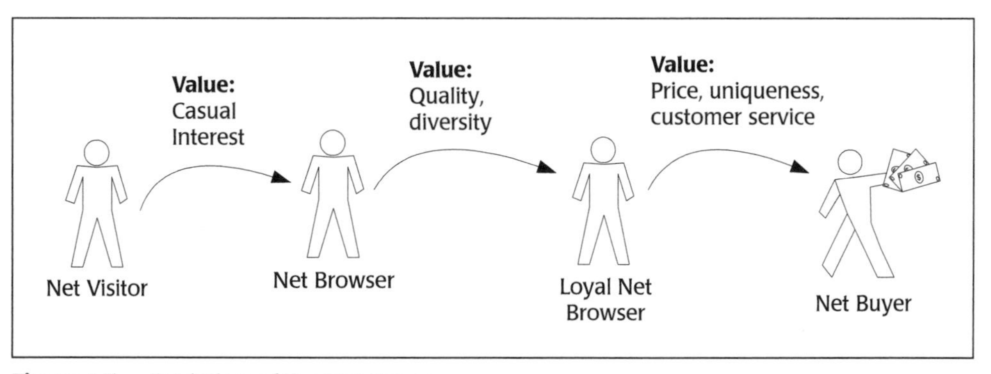

Figure 1.5 Evolution of the Net Buyer.

Summary

Killer content offers several key business benefits. First, high-quality content and services distinguish Web sites from others that offer information as a low-value commodity. Both Amazon and Cisco provide intuitive user experiences that accompany the sites' commerce orientation. As a result, both Web sites have gained strong marketshare and repeat customer bases. Second, value exchange that directly addresses site visitor needs provides incentive for the visitor to return. This loyalty offers the publisher a chance to convert first-time visitors into regular browsers, or casual browsers into paying customers. Third, killer content offers a mechanism to establish a relationship with end consumers. Using community features like chat rooms, bulletin boards, and e-mail, publishers can interact with Net users to learn about the types of content and products that prove most effective in the user experience and also enforce a positive impression of the site's brand. Killer content is an essential ingredient of any Web business model, regardless of Web site size.

But what does killer content look like? Who produces it? How can a publisher identify when content contributes to value exchange on a Web site? The next chapter addresses these questions by examining the role of relevance in the relationship between content and consumer.

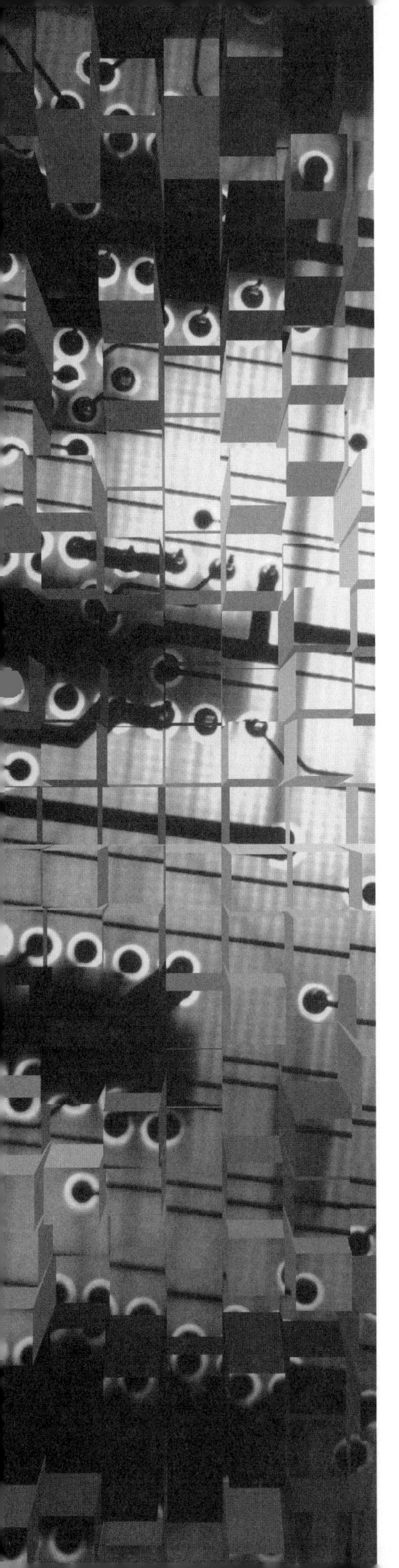

CHAPTER TWO

The Role of Relevance

Killer content is all around the Web. Once you understand what the value exchange looks like, killer content is easily distinguishable from pedestrian commodity content. *Commodity* content is information that is widely available and generally free to access on the Web. Large portal sites like Yahoo! offer commodity content (news headlines, stock quotes, sports scores) in order to drive additional revenue through advertising and e-commerce. Higher-quality content that also has "scarcity" (i.e., is not widely available on the Web) is considered *premium* content. Users generally access premium content through purchase or registration. Content providers, or publishers who create their own content, offer premium content that is unique and/or highly valuable to visitors. Premium content becomes killer content when the content enhances the Net user's interaction with the Web site.

There's a lot more commodity than there is quality on the Internet today, but a growing number of Web publishers are providing engaging and useful content to site visitors. This chapter examines the role of the user experience in killer content. Examples of value exchange that succeed and fail provide lessons on building content and services that benefit target consumers. First, let's take a look at free and premium content in one of the fastest-growing sectors of the Net economy: financial market services.

Free and Premium Content in Action

Jupiter Communications predicts that home investor households will increase to more than 20 million in 2003—a growth of 15 million households in just five years (see Figure 2.1).

Net browsers who manage assets online tend to be experienced Web users who use financial portals and sites to research investment decisions. One of the most popular financial Web sites today is TheStreet.com.

High Value Content Exchange: TheStreet.com

Led by Jim Cramer, financial pundit and hedge fund manager, TheStreet.com (*www.thestreet.com*) offers Net users financial news headlines, stock quotes, premium content, and other financial-oriented content and services.

First time users drop by TheStreet.com because of brand recognition or advertising. Voted as one of the top-ten investment sites by respected financial magazine *Barron's,* TheStreet provides timely commentary and trading advice on the latest ups and downs on Wall Street. TheStreet.com's content is also syndicated to high-traffic networks, such as America Online and Yahoo!, generalized news sites, and corporate intranets.

TheStreet.com's journalists are known for their shrewd analysis. For example, it was TheStreet journalist Adam Lashinsky who broke the story that

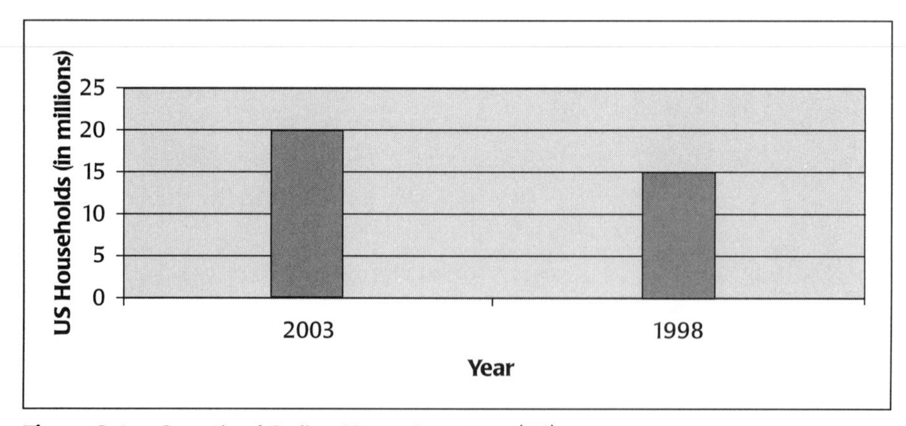

Figure 2.1 Growth of Online Home Investors (US).
Source: From "Financial Services Projections," Jupiter Communications Online Intelligence, New York, August 1999. Copyright © 1999 Jupiter Communications. Used with permission.

eventually forced online grocery Webvan to postpone its planned initial public offering (IPO) in the fall of 1999. Lashinsky discovered that Webvan was not observing the Securities Exchange Commission's required "quiet period" after announcing its IPO. The journalist called into a Webvan road show conference call and realized that Webvan was telling institutional investors information not covered in the company's prospectus (*Industry Standard,* 1999).

Among other predictions, Webvan executives were exhorting high margins in their grocery model (12% instead of the 4% of a typical grocery store) and some $300 million annual revenue from a single distribution center. Other major news services on and off the Web picked up the story. Within a couple of days, most financial news centers were either quoting or syndicating Lashinsky's news. This disclosure prompted the SEC to officially "raise concerns" that Webvan wasn't following the rules. Webvan postponed its IPO and began backpedaling rapidly. This kind of insightful journalism helps drive new visitors to the site, as evidenced by TheStreet's report that the average visitor spends twenty-two minutes on its Web site. That's almost nine times the number of minutes spent per visit at a portal Web site, according to the 15 August 1999 edition of "The Internet Capitalist" newsletter.

To increase visit frequency and user loyalty, TheStreet.com focuses on providing extensive up-to-date information about the day's market (an average of forty articles per day on financial topics) with about 30% of content freely accessible on the Web site (see Figure 2.2).

Premium Content

TheStreet.com depends on traffic generated by first-time and frequent browsers to drive advertising revenue, which by 1999 made up half the business's revenue. The other revenue stream came from subscriptions to the site's premium content services.

Subscription revenue volumes depend on TheStreet's ability to convert browsers to buyers. That conversion occurs when the content has high-enough relevance for the visitor to purchase a subscription. Net users purchase a monthly membership to premium content and services, giving them access to the content on the Web site reserved for subscribers.

Like other content publishers, TheStreet struggles to turn a profit on premium content (see Figure 2.3). With few Web sites outside of the pornography industry charging payment for content access, Net users are unfamiliar

Figure 2.2 Free Content on TheStreet.com.
Source: Copyright © 2000, TheStreet.com. Used with permission.

with the notion of paying premiums for information and services on the Internet. But TheStreet.com had managed to build a Net user base of more than 50,000 paying subscribers by the middle of 1999, with over half (65%) at the more expensive annual rate.

In addition, in 1999, TheStreet had a renewal rate of 90% for subscribers with annual memberships. The membership base is self-sustaining because of the utility and relevance of the services provided by the publisher. In a recent readership poll, subscribers reported that they made a trade more than ten times per quarter based on TheStreet analysis. In fact, in the first quarter of 1999, more than 300 articles exclusively published on the site had a negative or positive effect on the stock described. To bulk up membership and revenue coffers, TheStreet.com is turning its focus to different member acquisition strategies. One promising new technique is to partner with online brokers, such as Schwab, PricewaterhouseCoopers, and E*TRADE, who underwrite

Figure 2.3 Premium Subscription Content at TheStreet.com.
Source: Copyright © 2000, TheStreet.com. Used with permission.

subscriptions for their active brokerage account clients. Another technique is to provide premium content to corporate intranets, such as Fidelity's.

Lessons Learned from TheStreet.com

TheStreet.com demonstrates the diversity of value exchange on a single Web site. Daily financial news headlines and free quote lookups give first-time and repeat Net browsers incentive to use the site to keep abreast of the latest activities. For the serious investor, TheStreet provides a premium service that acts as a personal investment guide. Journalists weigh the pros and cons for particular stocks and provide a personal investor with informed commentary and advice for trading decisions. By addressing the needs of both browsers and buyers, TheStreet.com can leverage a hybrid revenue stream of advertising and commerce.

TheStreet.com Value Exchange

For Net Users: Financial trading news, commentary, and advice; free stock lookups; subscription premium service for exclusive articles and discussion groups.

For TheStreet.com: Traffic from a targeted demographic, sponsorships, syndication deals with news portals such as ABC.com, subscription revenue from site registrations, as well as bundled accounts with partner broker accounts.

Value Factors:

Credibility (respected journalistic opinions from pundits like Jim Cramer)
Innovation (exclusive in-depth analysis of stocks and funds)
Utility (information that a financial investor can immediately apply to the decision-making process)
Timeliness (up-to-date information about market activity)
Relevance (content that deals exclusively with financial information and services for a targeted demographic)

Retail Value Exchange

The notions of a flexible business model and a diverse value exchange apply to retail-oriented sites as well. Value exchange at retail sites mirrors the steps of a transaction in the physical world, with a different form of delivery (see Figure 2.4). The e-commerce merchant offers a desired product at a desirable price. The Net user provides payment for the product. The merchant's supply/distribution chain fulfills the order and ships the product to the buyer's address. The Net user gets his or her product; the merchant gets his or her money. In addition, the merchant may also get incremental advertisement revenue from the traffic that the Net user generated by visiting the Web site (or through selling information accumulated about Web site visitors).

The Web site acts as a storefront for the merchant's core competency, which is selling inventory through the Internet. A publisher adds value through forms of customer service—killer content, personalized product recommendations, customer service, reliable delivery, and an efficient return and

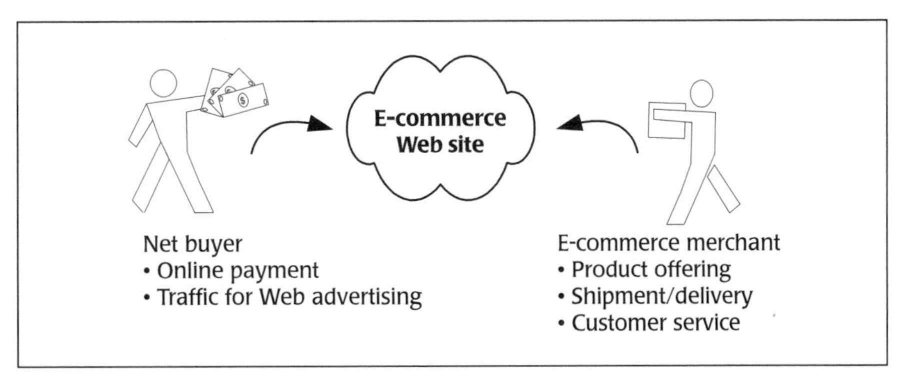

Figure 2.4 Value Exchange for Retail Goods.

refund process. Failing to build out any of those services means competing on a single dimension of value exchange: cost.

Perils of Single-Dimension Value Exchange

As journalist Matt Richtel of *The New York Times* phrased it, the Web is "the biggest mall, a discount shopping strip that stretches from Finland to Fresno." Competition can be ruthless among discount shopping strips. Egghead.com (*www.egghead.com*), reseller of PC hardware, software, and accessories online, found out just how tenuous a value exchange based solely on price can be. Egghead started out as a traditional retail store. After continuing losses in the mid-1990s, Egghead shuttered its remaining stores in 1997, moved to the Internet, and started selling about 40,000 products online, including computer hardware, software, and peripherals, as well as excess, reconditioned, and closeout brand-name computer products (Richtel, 1999).

In 1998, Egghead started suffering from online competitors who ruthlessly undercut pricing on similar inventory. For example, online competitor Buy.com sells inventory at cost; that is, Buy.com makes all their revenue from Web advertising on their site. In the fiscal year ending 3 April 1999, Egghead had $149 million in sales, a drop of 50% from the previous year. Whereas a company like Amazon focuses on services around inventory (such as recommendations, auctions, and Net-user-driven stores), Egghead concentrated instead on competing almost exclusively on price. As a result, Egghead lost out

Egghead.com Value Exchange

For Net Users: No clear value. Prices are at best equivalent and often higher than those offered by an online competitor such as Buy.com or even an offline computer retail superstore such as CompUSA.

For Egghead.com: Lower overhead in maintaining store inventory and storefront, direct sales channel.

Value Factor:

Utility (convenience of placing an order over the Internet)

to larger retailers and direct-sales companies. Even a boost in early 1999 as the featured software merchant on Microsoft's Internet portal MSN didn't help. In the summer of 1999, losses and diminishing returns forced Egghead to merge with online auctioneer Onsale.com in a $400 million deal.

Egghead's experience demonstrates the risks of operating on price alone. The margins are thin enough on the Internet without engaging in a Pyrrhic pricing war against other online retailers. After all, price comparison is evolving into a science. Intelligent agents offer comparative shopping services for Net users. An *intelligent agent,* or "bot," is a software program that performs an action on a Net user's behalf. At sites like mySimon (*www.mysimon.com*), the user types in the name of the product and waits as the bot crawls over available sites on the Internet looking for the lowest price (see Figure 2.5). The first generation of e-commerce bots just returned the price and location of the product on different Web sites. The second generation can also execute a simple task, such as locating the lowest cost on a product and collecting product and Web vendor reviews. This kind of automated and manual click-by-click comparison shopping undermines any attempt by the publisher to focus value exchange solely on low cost.

Diversifying Value Exchange

Canny Net vendors, such as CDNow, branch out with unique services in addition to low price point. CDNow (*www.cdnow.com*) realized the risks of operating purely on price and readjusted its offering. CDNow's primary revenue

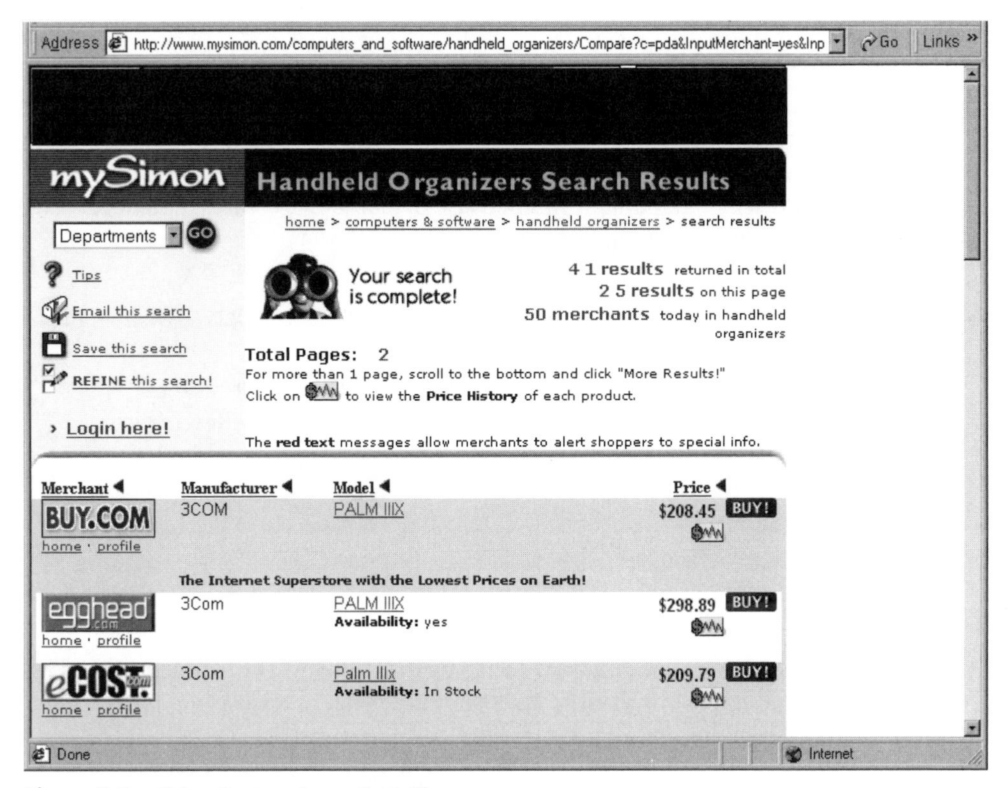

Figure 2.5 Price Comparison at mySimon.
Source: Copyright © 2000, mySimon.com. Used with permission.

comes from the sale of music albums online. Launched in August 1994, the company has built a customer base of 2 million Net buyers, a healthy brand name, and reported revenues of $56 million in 1998. But the music vendor lives in a very competitive space, occupied by Buy.com, Amazon, and a host of other large retailers. As a part of building out its brand and early revenue share, CDNow sets up extremely expensive deals with portals like AOL with a high customer acquisition cost ($45 per customer). When Amazon launched its music store in 1998 and promptly beat CDNow's profits in the first quarter of 1999 by 4 million, the music vendor took a step back and reevaluated its offering (Teleschi, 1999).

CDNow realized that Amazon's edge came from Amazon's loyal customer base. The music vendor saw that in order to compete with service-oriented

CDNow Value Exchange

For Net Users: Discounted prices on music albums, up-to-the-hour reports on the music industry, streaming audio previews for albums.

For CDNow: Higher traffic rates because of journalistic content, potentially higher revenue from purchases.

Value Factors:

Innovation (ability to "pre-hear" a song on an album through a streaming audio clip)
Relevance (in-depth information about artists and albums)
Utility (ability to purchase music while reading about the artist, or vice versa)
Credibility (journalistic quality of content adds a "music expert" ambiance to the site)

companies like Amazon, it had to add value beyond the price point of its products. In an interview with the *New York Times* in May 1999, Jason Olim, cofounder and chief executive of CDNow, described the decision to "go beyond just being a music store" by offering original content such as reviews and exclusive interviews with recording stars (Tedeschi, 1999).

The revised CDNow, says Olim, is "content, commerce, and community coming together." As part of the enhanced value proposition, CDNow provides hourly updates to music news and choice snips of music journalism, such as a previously unpublished interview with the late Kurt Cobain of the alternative rock band Nirvana. By establishing itself as a credible source of information on the music industry as well as a vendor of music, CDNow adds relevance to the browsing and purchasing experience for its customers. While it's too early to gauge the benefits in revenue terms, CDNow has at least established diversity as an incentive to visit the site.

Information Value Exchange

The sheer number of Web users indicates that browsers value the wealth of information online. In fact, a Jupiter/NFO survey showed that the three types of sites most frequently visited are e-mail, search engine, and research sites. Other popular sites involve news and local information (see Figure 2.6).

These common site choices emphasize Net users' predilection for using the Internet as a forum for information and services as well as shopping. Net users value the Web for the ability to find and use content that is relevant to their daily lives. There are two types of value exchange: one involving free content, in which a visitor uses a Web site for free information that is updated regularly (such as stock quotes or news headlines); and one involving premium services and content, in which the Net user must act in some way, usually by registering (for free) or making a purchase.

Like many large sites, Yahoo! employs free content value exchange in return for Web traffic. ESPN.go.com, a popular sports portal, uses both free content value exchange and premium content value exchange. ESPN's free content value exchange offers comprehensive sports headlines in return for Web traffic. Premium content subscriptions provide a supplementary revenue stream from sports fans who want more in-depth analysis of sporting events.

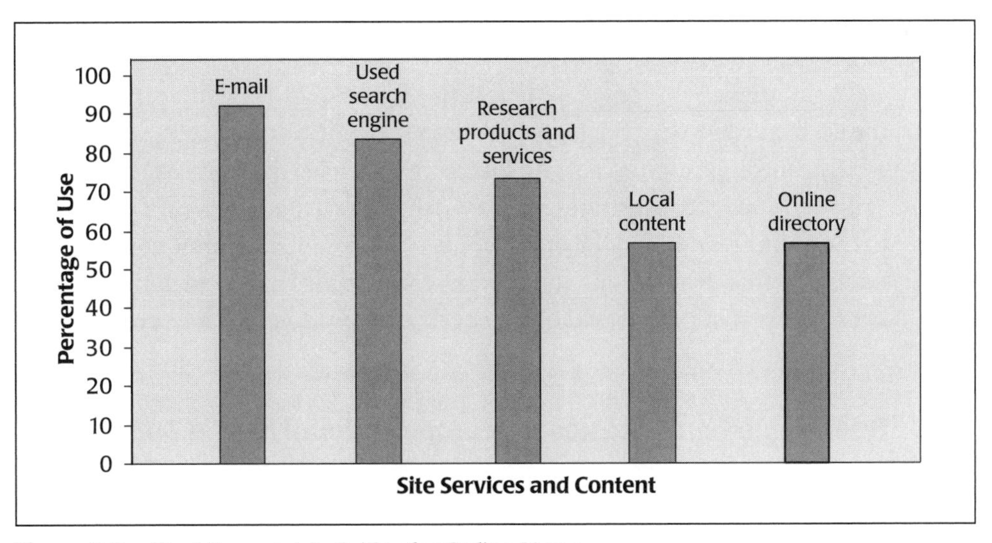

Figure 2.6 Most Frequent Activities for Online Users.

Source: Information from "Attitudes, Behaviors and Demographics of the Online User." The Jupiter/ NFO Consumer Survey (Figure IV.10), Spring 1999 ($N = 3{,}000$). Copyright © 1999 Jupiter Communications and NFO Interactive (pub. 7/99). Used with permission.

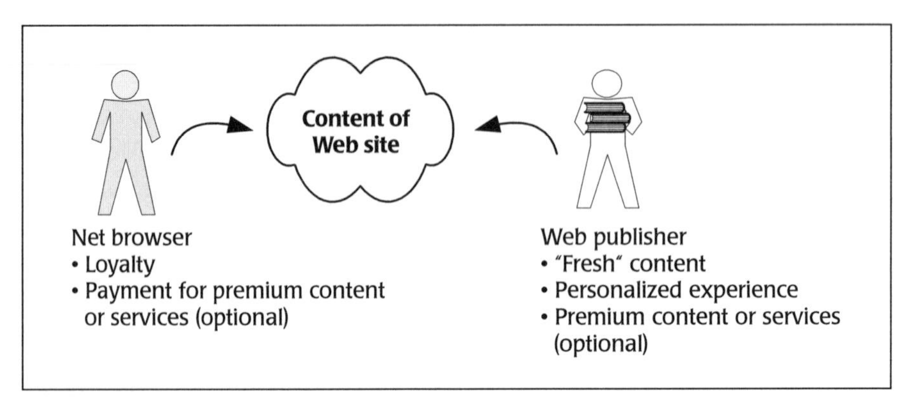

Figure 2.7 Value Exchange for Content Web Sites.

Value exchange that depends on Net user loyalty generally derives revenue from advertisements or sponsorships, while premium content can be used to pull in either traffic or revenue from subscriptions or other payment models (see Figure 2.7).

An offering of premium content and services looks very much like a retail goods exchange. The Web publisher offers high-value information or a service in exchange for purchase or membership. The Net browser provides payment or personal information for registration.

Value exchange around content depends on a host of factors ranging from the aesthetic (the quality of writing) to the technical (the speed and accuracy of searching). On one end of the spectrum, you have the *International Herald Tribune,* which demonstrates what to avoid in transferring a print magazine to the online forum. On the other end lies Salon.com, an online commentary magazine that provides highly relevant content for a targeted demographic. Let's look first at the low end of the spectrum for value exchange around content.

Poor Value Content Exchange: *International Herald Tribune*

The online version of the *International Herald Tribune* demonstrates how poor value exchange can damage a brand. Businessmen and the educated elite all over the world depend on the print *International Herald Tribune* for impartial coverage of international events. Unlike local or national newspapers, the *IHT* harbors no obligations to report on a single corner of the world or a single perspective in a localized conflict. As a result, the *IHT* has built a reputation with readership spanning Africa, Europe, and Asia.

Despite having one of the most respected brands in the print newspaper industry, the online version of the international newspaper in 1999 offers an unappealing presentation of the world's headlines. While the quality of the reporting cannot be disputed, the Web site offers little incentive for new and existing newspaper readers to locate and read articles. Visiting *www.iht.com* is at best uninspiring and at times frustrating. The *IHT*'s home page exudes a reluctant presence on the Web. Even the banner graphic is a poorly digitized image of a rolled up newspaper. The *International Herald Tribune* takes no pains to present information in a visually interesting or engaging way. The design and layout of the *IHT* site looks amateur in comparison to other news sites such as CNN.com. *IHT* organized its Web site almost exactly along print newspaper layout lines, paying no attention to the basics of browsing usability.

Digging down to an individual article requires clicking through sections and subsections. Key headlines lack abstracts to interest and inform the reader. Without abstracts, the reader does not have the information to effectively browse through a section. The reader has to read either minimalist headlines or the entire article in order to get a sense of the content. While the site offers search functionality, the results of the search are often inaccurate. The *IHT* doesn't even digitize photos to accompany text. The Net browser is left with columns of text with no accompanying visuals. In addition, the unnecessary graphics, such as buttons and banners, slows the initial page load to almost thirty seconds on a modem connection.

International Herald Tribune Value Exchange

For Net Users: Headline news and articles (without photos).

For *International Herald Tribune*: Online distribution of newspaper, but since the *IHT* does not advertise online, no ad revenue is generated through traffic.

Value Factor:

Innovation (the ability to offer online access to printed *IHT* news articles)

New Media "Cannibalization"

The *IHT* online Web site is more than an unlikely destination for news hounds to seek their daily information. The site actually detracts from the publisher's main source of revenue: the print newspaper. *Media cannibalism* in the online world refers to drawing Net users away from a commodity that generates revenue by providing the same value for free on the Internet. Every Web publisher who distributes information through printed media faces this risk in the Internet. If a Net user gets the same value for free online, he or she has little to no incentive to purchase the information or service.

Consider the following scenario. Let's say that I buy my *IHT* every morning along with my cup of coffee. I read the newspaper at my desk before beginning work. One day, I'm late and forget to buy the paper. I have some free time during lunch but don't feel like going out of the office to see if the newsstand has any editions left. Since I have Internet access through work, I navigate to the *International Herald Tribune* online. I realize that the newspaper that I pay for every day is essentially free online. I can just as well drink my coffee while reading the headlines on my computer. So I stop buying the print edition and simply read the online newspaper every day. If I'm a Net user who focuses on convenience, I may continue to frequent the *IHT* online simply because it's easier than searching for another source of news. But, it's more likely that I'm a Net user who wants to maximize the value of my browsing time. In that case, the *IHT* is in real danger of losing me as a customer—paying or otherwise—to another news Web site. Sites like CNN.com offer better usability, searching, personalization, and even e-mail delivery of the top headlines in preferred categories.

Media Cannibalization in the Retail World

Retail stores also face the risk of cannibalization. Companies that have strong mail order and/or retail presence, such as REI, risk losing consumers to their own online store. The company may lose the chance to up-sell, cross-sell, and otherwise promote products. REI addressed the issue of online media cannibalization by both increasing services around its products and integrating channels. For example, online kiosks in REI stores allow consumers to browse the online catalog and make orders while shopping in the retail store. The REI online Web site provides extra services around its product line, such as a bulletin board where visitors can post opinions about the pros and cons of outdoor gear (conveniently available for purchase a mouse click away). Both techniques encourage visitors to the online and physical stores to spend more time (and hopefully make impulse purchases) at the REI locations. By increasing services and integrating merchandizing techniques into varied channels, REI presents a strong and consistent brand presence both on- and offline.

Web publishers can avoid media cannibalism and the loss of customers entirely by requiring payment of information for accessing the information or by modifying the type of content in the online source. *The New York Times* online Web site leverages the value to the company of the digitized version of the newspaper by requiring membership to access the majority of the daily edition and to search the archives. To register as a *New York Times on the Web* (*www.nyt.com*) member, the reader provides key personal information (such as name and address) as well as optional demographic information. A data bank of thousands of readers provides *The New York Times* with a rich source of demographic information about existing and potential customers. The newspaper uses this information to effectively promote the print newspapers to members online or through mail. If a reader wants to read an archived article, he or she buys access to the article for $2.50 (the search for the article is free). Free access to the front page serves two purposes. First, the headlines pull Net users into the Web site so that *The New York Times* can generate revenue through online advertising. Second, readers can "preview" the content before handing over payment in the form of personal information or credit card.

The Wall Street Journal Interactive Edition (*www.interactive.wsj.com*) takes a similar tack to prevent media cannibalism of its printed version. It sells readers a daily edition for 75¢ and access to the online news for $59 a year (or only $29 a year for current print *WSJ* and *Barron's* subscribers). *WSJIE* also offers a two-week trial for the newspaper (in exchange for registration information) or promotes the 75¢ daily edition as a trial for the full-cost subscription. To differentiate an online subscription from a print subscription, the *WSJIE* offers additional services such as free searching through the archives of *Barron's,* a respected financial magazine.

Both *The New York Times on the Web* and *The Wall Street Journal Interactive Edition* avoid media cannibalization by adding unique value for Net users. The newspapers leverage the service benefits for the Internet channel. The Internet provides the ability to store archived articles and search for specific information. These two options are unique to the online versions of the newspapers. Scrolling through microfiche at the local library presents a much slower, less effective alternative. Both the archive and search capacities for information Web sites allow Net users to maximize the relevance of their user experience by quickly locating information of value. For example, Net users who are addicted to *The New York Times on the Web* crossword puzzle have the ability to buy access to crossword puzzles online, print them out, fill in the blanks, and return them to compare results against the online answers. Net users don't have to painstakingly cut out a tiny portion of a print newspaper, or remember to buy a newspaper with coffee the next morning to get the answers.

Bundling other related content can extend the value of an online publisher's offering, such as *The Wall Street Journal Interactive Edition*'s bundling of access to the *Barron's* investment Web site with an online subscription, as mentioned before.

Enhancing the value exchange around online content provides two key benefits for established publishing brands. First, the online newspaper serves as a separate revenue stream to augment printed distribution. Second, the publisher does not detract from the value of the printed version by giving away content for free on the Internet.

Building Brands through Value Exchange

Publishers who offer content only on the Internet don't have to worry about media cannibalism—there is no print media or physical store from which to steal customers. The term *pure-play brands* refers to companies that strictly use the Internet for distribution of content. Pure-play brands face different challenges in building a relationship with Net users, who have no context for the publisher. Net users have no familiarity with the source of the content and therefore no preconceived impression of the content's value (which can be a good or bad thing, depending on the circumstance). As a result, pure-play publishers have an even greater need than established brands to provide a highly relevant experience for users. Let's take a look at how online magazine Salon uses innovation and quality of content as the basis for value exchange.

The success of Salon online magazine depends on providing a highly relevant experience for its Net user base. Since Salon.com exists only on the Internet, the site focuses on its special blend of quality journalism and commentary to build brand awareness. Net pure-play companies face a bigger challenge in gaining first time visitors to their sites. Established brands like the *New York Times* use their instant name recognition with online users as the basis for a value exchange. Salon depends on its reputation among Net users and the distribution of content through deals with news portals to spread the word.

Based in San Francisco, Salon offers a combination of commentary on political, social, and popular culture issues by well-known literary talents such as Garrison Keillor. Although not a local and international news center, Salon is building a reputation for quality investigative and specialized reporting. During the 1999 Monica Lewinsky scandal, it was Salon that first broke the story about the adulterous affair carried on by Congressman Henry Hyde, chairman of the Judiciary Committee considering President Clinton's impeachment case. Salon's Travel section is so respected that in 1998, the magazine won two Internet travel journalism awards (*Best Internet Publication/Web Site* and *Best Internet Travel Article*).

While commentary offers the core of the value exchange with Net users, Salon also offers a series of services for Net browsers. A section called "Table Talk" lets readers interact with Salon writers through e-mail and posted letters. In addition, Salon gives visitors the option to register with the Web site to get e-commerce discounts from Net vendors such as Barnes&Noble, to join a members' discussion lounge, and to use free e-mail. To build community, Salon provides a subscription-based discussion group (available through the 1999 acquisition of Web community vendor The WELL) that focuses on issues raised on the Web site.

Salon's readership typically consists of highly educated professionals who are loyal and frequent visitors to the Web site. The plethora of services offered by the publisher provides loyal readers with a variety of different options. As a result, the target demographic stays longer during a single browsing session, stopping by discussion groups, chats with editors, polls, and entering sweepstakes. Not surprisingly, Salon's revenue depends on advertisements and sponsorships. Salon also has content distribution deals with portals and news sources such as Lycos, CNET, and REUTERS to increase the visibility of the magazine's journalistic content. Salon also gains some revenue from affiliate programs with partner Net vendors. (The publisher collects between 4% and 6% on transactions at bn.com that originate at the content site.)

Salon.com Value Exchange

For Net Users: Exclusive commentary and editorial content by well-known literary figures. Interactive communication around topics with other readers, the writers, or site editors.

For Salon.com: Highly targeted demographic of educated, professional Net users. Users are active members of the community and tend to participate in polls and surveys that ask for opinions (after all, it is an editorial Web site). Revenue streams include advertising, affiliate links with retail vendors, sponsorships, and content distribution through news portals.

Value Factors:

Innovation (editorial articles and commentary on often sensitive social, political, and cultural issues)
Credibility (reputation of insightful editorial information and reporting)
Relevance (highbrow topics speak to the target demographic of young, Internet-savvy educated professionals)

Salon.com and the *International Herald Tribune* online represent two ends of a content value spectrum. Salon provides highly relevant content to a targeted demographic. The *IHT* simply offers a digitized version of the print newspaper, without regard to usability features such as effective searching of photos. Salon's emphasis on reader communication with Salon writers and editors builds a sense of community around the Web site. The *IHT* doesn't even provide a way to contact customer support for online or print newspaper delivery problems. Ironically, while the *IHT* is recognized around the world as a respected publisher brand, it is the Net-based Salon that has built a reputation among Net users as a source for quality. Web publishers of both established and new brands on the Internet succeed according to the level of value and relevance offered to Net browsers.

Summary

Understanding the importance of killer content requires an understanding of the types of value exchange around content. This chapter has introduced the notion of two general categories of value exchange: free content and premium content. Publishers offer free content value exchange in order to attract and keep Web traffic. In free content value exchange, a publisher provides up-to-date information and services for site visitors. Often, the frequency of the updates and uniqueness of the content or service provide Web traffic for advertisement revenue. For example, Yahoo! aggregates a variety of different types of content for consumers in order to support advertisement rates and also drive e-commerce transactions on affiliated shopping sites. E-commerce-oriented sites such as CDNow use free content exchange as an added layer of service to their retail environment. As sites like Egghead.com prove, offering a sterile, single-dimension experience to consumers spells a low barrier to entry for competitors and means a difficult road in a highly competitive space.

Premium content exchange requires either a payment or free registration (in which the site visitor "pays" with personal information). Sites that offer highly relevant and unique information, such as TheStreet.com, use premium content exchange as one of several revenue streams. A small but growing number of content providers use premium content exchange as an essentially supplementary revenue stream for advertising and other business models.

The choice of free or premium content exchange defines the relationship between a Web site and a Net visitor. This relationship also changes by Web site objective. A Web site that displays corporate information offers a very different type of free content exchange than the type offered by the user experience at an entertainment-based Web site. The next chapter discusses the categories and critical factors of value exchange based on the objective of the site and the subsequent user experience. Publishers and users of Web sites can study the variations of value exchange in order to define personal preferences, review other publishers' mistakes and successes, and determine if their own site visitors' goals and expectations are being met.

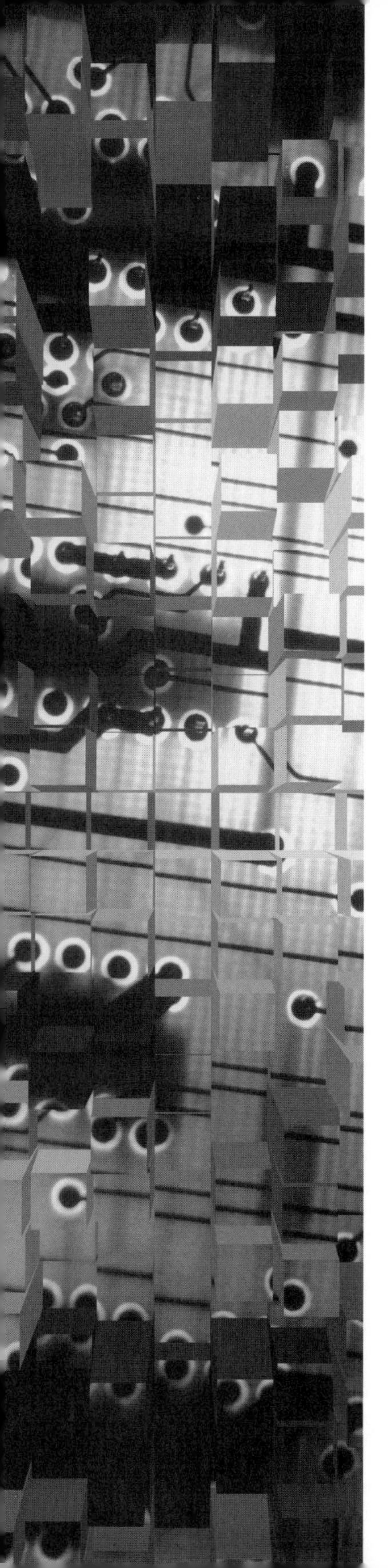

CHAPTER THREE

Value Exchange Variations

A single type of value exchange can not be applied universally across the Net. Services offered by one type of Net publisher might not prove effective at another site. The value exchange at a corporate Web site is very different from the relationship between a buyer and a seller at the online auction site eBay. In an eBay transaction, the buyer and the seller have the same objectives (albeit at different price points) for the transfer of ownership of a product. eBay's core competency is in providing the marketplace for Net-user-to-Net-user buying and selling. The value exchange that makes the site so successful lies in the additional services around the experience, such as escrow payment options and Net-user buyer/seller rating services.[1] eBay's enhanced services help facilitate the transaction and make a community out of the auction environment. Those same services would sit strangely at a corporate site, where the visitor's goal is to find more information about the company or product. This chapter investigates how a Web site's orientation impacts the evolution of services and commerce in different value exchange categories.

[1]At eBay, buyers and sellers can provide feedback on their experiences with particular individuals. This self-regulation tactic helps warn potential auction participants if a bidder or buyer is prone to "flaky" behavior (such as backing out of an accepted bid or not delivering purchased goods).

Categories of Value Exchange

Value exchange categories provide the framework and context for figuring out what services fit best with your particular Web site. Content and services at Web sites evolve constantly. Amazon started out as a bookseller. Try to categorize the Amazon model now, and you end up with a long, hyphenated descriptor that strings together words like bookseller, auction house, and merchant host. Many vertical portals (Web sites that provide aggregated information about a single subject) began with pure content but have started integrating commerce in the form of product referrals to retailers. These evolutionary paths involve different variations of value exchange. We will take a look at four categories of value exchange and examine the services involved in each.

The four categories of value exchange are as follows:

1. *Promotional,* which provides single-topic information about a particular product or company on the Web in exchange for brand recognition;
2. *Commerce,* which offers inventory (physical goods, digital goods, or digital services) in exchange for payment;
3. *Content,* which offers aggregated information (general news or commentary) in exchange for site traffic to drive advertising; and
4. *Entertainment,* which provides rich media niche content in exchange for site traffic to drive advertising or cross-site referrals.

Most Web sites offer at least one of these categories, depending on the content and the goals of the publisher. Take, for example, the online destination for popular news provider CNN (*www.cnn.com*). CNN's strong Internet offering of current news and events is based on the concept that all information needs—online with the Web site and offline with the cable channel—are best met by a single provider. CNN's value exchange involves both *content* (information in exchange for Web traffic) and *promotion* (information in exchange for brand recognition).

Entertainment portal Yahoo! Broadcast entices Net users by regularly broadcasting popular sporting events or earnings announcements, and then attempts to hold their interest with offers of related albums, books, and other commodities. Yahoo! Broadcast depends on an *entertainment* value exchange (an innovative experience in exchange for Web traffic) in order to fuel *com-*

merce transactions. Consulting companies like Razorfish use the Web for promotional lead generation, offering contact information, customer lists, and other sales-related data.

At the core of each category of value exchange lie two factors: the Net user's *goals* and the Net user's *expectations*. Value exchange depends on the Net publisher and Net user working from a common set of goals and expectations. This maxim applies to any type of value exchange, whether it involves an online transaction or not. The publisher absolutely must understand the goals of the site visitors and provide what visitors come to the site to find, whether it's information or inventory. If the publisher doesn't understand why a Net user makes a visit, he or she cannot turn a casual visitor into a Net browser or buyer and retain that individual's loyalty. (For example, the producers of the online *International Herald Tribune's* awkward and confusing navigation impedes the Net visitor's goal of finding information.)

Publishers must also understand the Net user's expectations and offer the user experience that visitors anticipate. The Net visitor arrives at the Web site with expectations set by advertising, word-of-mouth recommendations, or previous experience. Failing to meet, or at least come close, to those expectations results in lost traffic and sales.

The Web site of consulting company Arthur Andersen may satisfy the Net user's goal of information, but it does not present an entirely effective calling card (*www.arthurandersen.com*). Arthur Andersen, an international consulting company with over 70,000 employees, provides a Web site for potential clients to find out more about the firm. The Web site contains an impressive array of information, especially regarding the consulting giant's methodology. Arthur Andersen understands that potential clients' goal is to find out as much information as possible about the services that the company provides.

But Arthur Andersen bogs down in providing a positive user experience because of poor site design. The performance of the Web site is deadly slow and directly attributable to the number of unnecessary graphics and controls on every page. Even over a fast integrated services digital network (ISDN) connection, simply loading the site's home page (complete with glitzy entry) takes over twelve seconds.[2] The performance of the home page is probably

[2]Clocked in October 1999. Performance may change as the company tunes its Web site.

related to the use of Lotus Domino 5.0 (a Web server not known for its performance), a considerable amount of JavaScript, and about fifteen image files (some as big as 20 KBs).

In addition to poor page performance, Arthur Andersen buries its client list so deeply in content that it requires a search to locate customer names—and even then the lists are partial, by industry or by geography. The home page doesn't offer customer testimonials, case studies, or downloadable information files. Printable brochures and other product information lie several layers deep on the Web site. Subsections of the Web site differ radically from the site design, switching color schemes and navigation patterns. After multiple frames in which menu navigation is tucked in the left frame, the site suddenly shifts to clickable graphics for navigation.

Because of poor screen performance, uneven navigation, and lack of downloadable information, the Arthur Andersen Web site doesn't meet the Net user's expectation of finding information about the company's services quickly and easily. In addition, the poor site design does not endorse the company's ability to work within an Internet economy. Arthur Andersen runs the risk of losing potential clients who either can't locate information or are put off by the poor site design as an example of the consulting expertise. The Web site fails at value exchange by not successfully delivering on both the Net visitor's goals and expectations. The importance of fulfilling Net user goals and expectations remains constant throughout all four types of value exchange.

Promotional Value Exchange

Value exchange on a promotional Web site consists of providing information about a company or product to Net visitors quickly and easily. The bulk of Net users involved in a promotional exchange are the audience of an advertising campaign and/or potential clients of a company. That is, the majority of these visitors have arrived at the Web site with specific intent, for the content. They might be curious about a funny television commercial for the company or a positive review in a newspaper article, or actively in search of contact information to make a deal.

The publisher's benefit in a promotional value exchange revolves entirely around the advancement of a brand. That advancement takes shape in either an *advertising* context (extending a marketing message to an online venue) or a *sales* context (demonstrating expertise for a service or product). Advertising context provides the Web publisher with the opportunity to maintain the promotional themes for an advertising campaign in print or online media. Sales context allows the publisher to promote the company's core product or competency. Most sites combine the two contexts, as illustrated by the corporate site of "strategic digital communications" provider Razorfish.[3] Razorfish's site handles a promotional value exchange that satisfies both the goals and expectations of Net visitors who want to find out more about the Web site design firm. Razorfish uses the corporate Web site design to advertise its own expertise, and provides extensive contact information to encourage sales.

Visitors to Razorfish's corporate site expect a high standard of usability and design innovation. Being a Web design shop, the company cannot fail to meet these expectations without damaging its reputation (and brand) and losing potential business. Razorfish's Web site attracts potential clients and curious visitors because of its reputation as a leader in graphic innovation and Web site strategy. The firm uses the Web site as a central repository for corporate information, sales contact information, a portfolio of work, and links to new innovations in graphic design. Razorfish helped build its reputation in the Internet industry by also using the corporate site as a display case for media and presentation.

The company, founded in 1995 by Jeff Dachis and Craig Kanarick, started off in the founders' proverbial garage, which in New York City takes the form of a small apartment. The company's big break came in 1997, when Charles Schwab & Co. Inc. hired the small Web design firm to restructure its corporate Web site. Since then, Razorfish has branched out from Web design to e-commerce system infrastructure with the acquisition of i-Cube, a company that converts legacy infrastructure for e-commerce. Clients include eBay, heglobe.com, Time Warner, the Smithsonian Institution, and Ericsson.

Established brands like Price Waterhouse retain Razorfish's services because of its reputation for building effective navigation in a well-structured Web site. Young Internet companies—such as game site Bunko!—use Razorfish

---

[3]For an example of promotional value exchange that consists purely of advertising, check out the Web sites for any of the major soft drink manufacturers such as Seven Up Corporation (*www.7up.com*). These companies use the Internet to extend a television or print campaign to an online Web site.

because of its innovation in graphics and digital media. Razorfish leverages the layout of the corporate Web site (*www.razorfish.com*) to provide value exchange for both types of potential clients.

Its Web site focuses on providing Net users with information on Razorfish "solutions"—company mission, people, partners, and sales information. Most corporate Web sites stop with this kind of packaged sales pitch to the Net visitor. Razorfish takes the value exchange a step further by giving site visitors additional resources to explore the firm's design work. On the corporate site, there's an online portfolio of work for dozens of clients. Potential customers can browse through various work, searching by industry. The site also contains practical information for decision makers: clean presentation (advertising by example), corporate history, client lists, real-world portfolio, and contact information.

Razorfish does not address expectations around innovation directly on the corporate site. Rather, the "dope stuff," as termed by Razorfish, lives on two Web sites linked to the corporate location. Razorfish Studios and RSUB are divisions of Razorfish and act as showcases for the firm's design talent as well as destination sites for other Web designers interested in new media presentation.

The Razorfish strategy for promotional value exchange illustrates how added services make the relationship between Net visitor and Web site more engaging. What these enhanced services are depends on the nature of the company or product. Remember that Razorfish's business is Web design. It's a natural extension of the firm's business to offer an online portfolio and Web studio as part of the corporate statement. It would be neither practical nor particularly effective for Arthur Andersen to do the same with their Web site. Arthur Andersen specializes in improving business processes. An effective value exchange for Arthur Andersen might include organizing and presenting information in a way that allows the Net visitor to find answers quickly—giving the impression of organization and thoroughness. The equivalent of the online portfolio for Arthur Andersen would be case studies and statistics of how former clients benefited from consulting.

Ultimately, the benefits of a promotional value exchange lie in brand awareness and (potentially) sales driven by the Web site. The *key element* to promotional value exchange is effective distribution of advertising or marketing message.

Promotional value exchange revolves around two factors: the publisher's goal of getting visitors to buy the product or service offline (in a bricks-and-mortar store), and the Net user's goal of obtaining information in order to make that decision. The Net user's expectations are closely tied with the product or service's reputation, which is established through direct experience or advertising. If the brand represents expertise in a given area, the visitor expects that the site will present content in a professional and attractive manner.

There's no online business model around promotional value exchange. *Promotional value exchange* is about advertising a core product or competency to drive sales offline. You can't sell advertising space for other companies on a Web site because it distracts the visitor's attention from *your* message. Promotional content itself has little commercial value. A corporate or advertising site may sell brand commodities (Razorfish sells prints of original art, for example), but the company certainly doesn't base its revenue numbers on the proceeds.

Conversion goals for a promotional value exchange fall second to the publisher's concern with getting a message across to the Net visitor effectively. Conversion of casual visitors to loyal users may be *related* to the effectiveness of the site but do not necessarily constitute the core of the value exchange, as is the case in the other categories. For other types of value exchange, conversion ratios are crucial to the success of the relationship. They are most important when dealing with commerce transactions.

Commerce Value Exchange

At its most basic, a commerce value exchange provides a product in exchange for some form of payment. Few retailers do commerce value exchange well. The fact that three-fourths of online users have never made an online purchase (see the "Net Browsers and Net Buyers" section in chapter 1 for more detail) illustrates both the enormous opportunity and the challenge in building an effective commerce value exchange. The Achilles heel of commerce value exchange is the lack of enhanced services beyond the purchase transaction. A purchase must be more than simply paying for a product. A purchase may involve selecting a product, filling out the order form(s), receiving the purchased item, and potentially contacting customer support for help. Each of these elements in a purchase constitutes a service in itself. The publisher who offers commerce value exchange without excelling at these services ends up losing customers to the Web publishers who do.

A single dimension of service in a promotional value exchange is acceptable, as long as that service addresses user expectations and goals. For example, Razorfish's value exchange provides only one service: access to information about the firm's history, talents, and client list. In contrast, limiting services for commerce exchanges spells inevitable problems. Competitors who could afford to sell the same inventory at literally no margin eventually forced Egghead, who competed solely on the basis of cost, into consolidation. If the publisher depends solely on a single dimension to the value exchange, such as price or the uniqueness of a business model, the publisher risks irrevocably losing the market advantage when that dimension is inevitably challenged. (We'll see how that plays out later in this section, in an examination of the travel site Priceline.com's difficult position.)

Net User Goals and Expectations

Net users have the highest expectations around commerce of all the categories of value exchange. Net users expect value in more than price point, unless the price is so incredibly low that the savings obliterate other concerns. In 1999, Forrester Research, an analyst firm based in Boston, published a report (Morrisette, 1999) that examined purchasing and browsing patterns for Net users. Analysts looked at what motivated Net users to become Net shoppers, and what drew Net shoppers back to a Web site for repeat purchases. Based on shopping statistics from 1998, the report strongly advises Web publishers to "differentiate on service and on-line experience—or be the low-cost leaders" in order to succeed as Net vendors. It takes more than a single dimension of service to engage a Net user in a sustainable commerce exchange relationship.

The Net user's goals and expectations depend on what the Net user values about the product (for example, prompt customer service).[4] As discussed in the first two chapters, the key to Amazon's popularity among Net users lies in the retailer's customer service, Net user profiling, and selection. Amazon makes up for thin margins on retail goods by focusing on high volume throughput for an efficient supply–distribution network. Without enhanced

[4]Forrester Research estimates that by 2002, 70% of all online shoppers will make purchases based on the *value* of the online purchase rather than the convenience of buying online.

services that bring in repeat customers, publishers can't survive the thin margins on the Web.

The Dangers of Single-Dimension Value Exchange

The online travel industry is another market that depends on volume to turn profits. The travel industry composes the largest segment in online retail market, more than books, music, or electronics. Jupiter Communications (*www.jup.com*) estimates that the online travel segment will do $4.2 billion in business in 1999 and $16.6 billion by 2003. That's with only 2% of the world's travel reservations booked online (Shore et al., 1999).

Despite the projected revenues, there's not a lot of room for Net travel vendors. The margins are too low. The only merchants who succeed are the ones that deal in volume. Consider the commissions on bookings sold online. A travel agency in the real world receives 5% to 10% of a transaction, which, on a $500 booking, can mean as much as $50 for services rendered. The online commission collected by a Net travel vendor goes as low as 2%, or $10 on that same $500-dollar ticket. Net travel vendors then depend on volume to make up the difference and cover the costs of maintaining the mammoth infrastructure required for an online travel-booking site. As a result, the online travel industry has seen even large industry players turn to consolidation in order to succeed. Travel giants Travelocity and Preview merged, creating what Terrell Jones, president and CEO of the new company, estimates is the third-largest e-commerce site in the world, behind eBay and Amazon.com (Shuchman, 1999). With major travel sites pooling resources to support a business that depends solely on cost, other travel services struggle to provide additional services that round out the core exchange of bookings for payment.

The Problem with Priceline.com

Travel vendor Priceline.com thought that its "name-your-price" business model provided a unique service that other competitors couldn't match. In fact, Priceline.com successfully filed a patent on its reverse auction model—the ability to make a bid on commodities as they become available. Priceline.com sells travel bookings, cars, home refinancing, and home equity loans by finding products that match a Net user's price. Jay Walker, who founded the company in 1997, realized that there was a significant business opportunity in the fact that each day major airlines have millions of empty seats. Walker's

Wallet Digital Brain trust developed a "name-your-price" auction system that forms the basis of all aspects of the Priceline.com service. The airline component was launched in 1998 and then expanded into hotel reservations and car-buying services.

Net users can fulfill the goal of obtaining a cheap plane ticket at Priceline.com, although any expectations of Web site usability are quickly dashed when going through the order placement service. The forms are long, and the site throws in the occasional unsolicited pitch for a partner vendor as a part of the process. Once the Net user wades through order placement, the user then authorizes Priceline.com to buy the ticket if the price becomes available (see Figure 3.1). The user has no control over the airline choice and must simply accept a ticket based on the correspondence between price and preferable travel time. Priceline.com accepts Net user bids if the price is no lower than

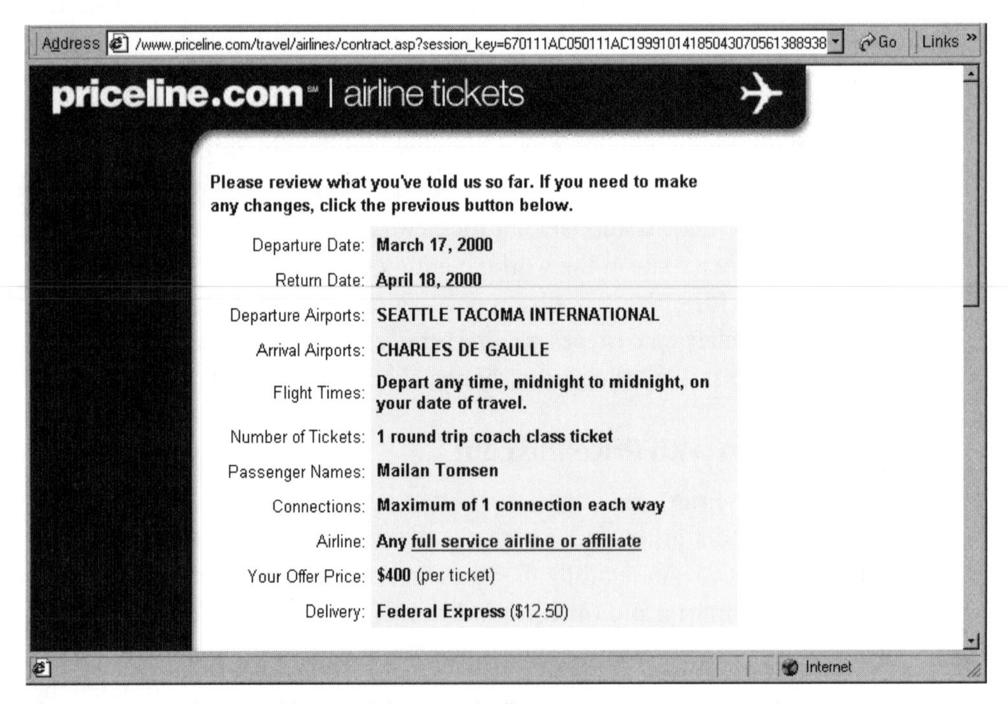

Figure 3.1 Making a Bid on a Ticket at Priceline.com.
Source: Copyright © 2000, Priceline.com. Used with permission.

30% of the lowest fare for the route. The company makes its money on the margin, by pocketing the difference between the price paid by the individual and what the company pays for the commodity.

The problem with the site's value exchange is very similar to Egghead's situation. Priceline.com bases its value exchange on a single dimension—its reverse auction model. Granted, that's a pretty nice deal, and it's been the stimulus for Priceline.com's revenues ($161 million in the first six months of 1999) and customers (more than 16 million by 1999). But the site's business—travel booking, cars, mortgages—depends on the uniqueness of the business model. Priceline.com doesn't even offer minimal service enhancements around the actual bidding process, such as descriptions of travel destinations. The site clearly revolves around the Net user's goal of finding a cheap price without addressing any expectations around service.

Priceline.com's complacency stems from its broad patent on the reverse auction model. Take away the patent and the vendor's entire business is threatened because of the lack of diversity in its value exchange. Within fifteen months of launching the service, Priceline.com faced two significant challenges to its patent. One came from Microsoft's travel site Expedia, which—despite the patent—added a hotel reservation service based on reverse auction bidding. Microsoft spokesman Tom Pilla brushed off concerns about violations of the patent in an October 1999 interview with the *E-Commerce Times.* "We certainly respect the intellectual property of others," said Pilla, ". . . [but] this is a desperate attempt by Priceline.com to avoid competing with Expedia on the merits" (Conlin, 1999).

The other more serious challenge came from a lawsuit by patent lawyer Thomas G Woolston, "who filed a patent application describing an electronic market for used and collectible goods that competes with Priceline.com's . . . patent" (Byrne, 1999). Woolston argues that he beat Priceline.com to patent by sixteen months and intends to license "his" issued patent to the owner of a competing Internet travel service. If Woolston wins (or other services simply adopt the model as Expedia has done), Priceline.com suffers more than bad publicity. Its entire value exchange depends on a novel way to obtain the lowest price. Once the novelty is removed, the only thing left for Priceline.com to compete on is price and service. With the low margins in the travel industry, undercutting margins further is next to impossible, and Priceline.com is significantly behind sites like Expedia in services.

Priceline.com admitted that "if Woolston prevailed in both an interference and an infringement action, then Priceline.com could be enjoined from conducting business through the Priceline.com service to the extent covered by the patent claims awarded to Woolston" (Byrne, 1999). Take that patent away and the site will have little left in its value exchange. The Priceline.com case demonstrates how essential it is to provide additional services beyond a single cost-related dimension in the value exchange.

Content Value Exchange

Services are equally important for content value exchange, which also depends on high volumes for success. Content value exchange revolves around information and services that represent, as the inventor of the World Wide Web Tim Berners-Lee put it, "the primary embodiment of the ways in which we work and play and socialize" through the Internet (Berners-Lee, 1998).

The Net visitor is looking for content that educates and informs. (Entertainment has evolved into a significant enough area of Web usage that it gets its own discussion later in this chapter.) These expectations hinge on the following value factors: *relevance* of the content to the Net user, *timeliness* of the content, *innovation* in presentation, *credibility* of the source, and *utility* in finding and interacting with content. Content providers specialize in addressing one or more of these value factors.

Content takes many forms, ranging from the copy of political commentary to the streaming media of online radio broadcasts. Publishers produce content (copy, graphics, rich media) or syndicate content from content providers. Content can be distributed by its owner (the content provider) or licensed and syndicated to an aggregator of content (a portal or intranet). Most commercial Web sites engage in some form of value exchange around content. Media sites such as commentary Webzine Salon.com and general news portal CNN Interactive revolve entirely around the exchange of information for Web traffic.

The simple exchange of information for Web traffic has proven unprofitable for all but the most highly trafficked destinations. Commerce on the Web is evolving beyond retail transactions into new types of value exchange. The next chapter discusses these new models, which include syndication, pay per access, and subscription, and talks about how they open up new opportunities for earning revenue. Salon.com readers, for example, use the site to access social, political, and cultural commentary and investigative reporting. As the creator of the content, Salon owns the rights to its distribution and syndicates editorials and commentary to other Web sites. Networks, portals, and other syndicators depend on the content from content providers like Salon.com to enhance the value factors in their Web sites.

Other Web sites, such as portal Yahoo! and travel booker Travelocity, use content value exchange as an additional service for Net users to drive revenue from ticket sales. Yahoo! syndicates content from a variety of sources, including TheStreet.com and Salon, in order to retain Net visitors on the Yahoo! site(s). For example, Yahoo!'s personalization service, MyYahoo!, lets Net users pick the types of syndicated content to fill their customized news page.

Commerce sites also use content value exchange to drive commerce sales. Travel booker Travelocity.com offers a Travel Guide, with entries on every country in the world. Each entry features an overview about the place, key facts, how to travel, and when to go. The Travel Guide's content has a singular purpose: to provide information about different vacation spots, which, in turn, powers airline ticket and hotel reservation sales. Fitness and health portal FitForAll uses what founder Marvin Chow calls "passive content" to attract visitors to the site. Passive content includes general interest articles on the latest issues in health and fitness. According to Chow, the fitness and health portal helps users "achieve a higher quality of life" through exercise advice, product recommendations, and online personal trainer services (phone interview, 15 Aug. 1999). FitForAll uses content to accommodate three tiers of users: casual browsers, who have access to general interest articles on health issues; members, who register for the free personalization service and set up a personalized home page with syndicated news and services of their choice; and consumers, who buy premium membership at $8 a month for access to personal trainer services via e-mail and bulletin boards.

FitForAll obtains most of its content from content providers, such as Vitamins.com and Healthnotes.com, which focus more narrowly on health-related issues. This syndication model lets the start-up fitness portal provide high-quality content to visitors without spending the resources for content development. Instead, FitForAll provides a centralized location for consumers who want to learn more about general health and fitness. This emphasis on education for personal fulfillment addresses the goals and expectations for users visiting the portal.

Networks and Portals

Portals provide an entrance to other information or services. On the Web, portals come in many different varieties, including access to an extremely limited number of networks or a larger number of "vortals" (vertical portals). These content networks come as either a loosely affiliated combination of sites (like the GO Network) or a variety of services under a single domain (like Yahoo! and Excite). The GO Network (*www.go.com*) is an organized affiliation of major Web properties, such as ABC.com, Disney.com, ESPN.go.com, and Mr. Showbiz.com. The network, which resulted from a merger of InfoSeek and Disney Corporation, provides the excellent content of its property sites as well as portal services such as free e-mail, Net user Web pages, personalized news, local entertainment listings, chat rooms, instant messaging, and directory and search services. Proprietary network America Online resembles the GO Network in that it offers content and shopping to subscribers through other content providers.

Yahoo! represents the other type of network, which consists of multiple service sites within a single branded domain. Net users gravitate toward these networks based on the convenience offered by multiple types of content available in a single location. In addition, the high-traffic networks and portals provide reassuringly familiar brands. (Yahoo! has one of the strongest Web brands.) The network gets well over a million hits a day from users all over the world. The main Yahoo! Web site (*www.yahoo.com*) is the most popular Web site in Japan and the dominant Web portal brand in Europe. It's a long way from where Yahoo! began.

Evolution of Network Content Value Exchange

Yahoo! started out in 1995 as a search engine, providing information (links) to Net users who wanted to locate sites by keyword. Within four years, Yahoo! evolved into a network of syndicated content and services on specialized sites, such as Yahoo! Financial and Yahoo! Auctions. Yahoo!'s personalization service allows a Net user to set up a customized Web page with preferred categories of all the content in the Yahoo! network. For Net users who want greater depth on a particular subject, Yahoo! offers specialized Web sites that aggregate content from different sources. The network also covers commerce value exchange through classifieds, auctions, and shopping links.

Convenience and brand recognition form the basis of a network's content value exchange. The Net user's goal of obtaining information, shopping, or simply browsing is addressed somewhere within the network. The user's primary expectations are *usability* (the ability to navigate to the desired content) and *credibility* (the aggregation of quality content under the single brand). The successes—even the evolution—of these sites are limited by the economics of Web traffic and content aggregation. Like Net travel vendors, networks depend on site traffic for most if not all revenue. These sites get huge volumes of traffic a day (in the first quarter of 1999, Yahoo! reported 47 million registered users and 235 million page views). There are not many other sites that can boast such a volume.

In addition, the value that these networks provide (convenience and credibility) comes at a high cost for marketing and content acquisition. Witness the hundreds of millions of dollars Yahoo! spent in acquiring properties for its network. Only a few players in the Internet economy have the kind of capital for the mergers, acquisitions, and/or other deals required to build out the variety of content and services available at Yahoo!.

Vertical Portals (Vortals)

Vertical portals, or vortals, on the other hand, can focus on a single subject for content and services on their sites. A vortal provides content and services around a single topic. Users visit a vortal to obtain in-depth information about a particular topic in one location. The content itself may be produced by the publisher or syndicated to the Web site from different sources. Value exchange for vortals succeeds when the target demographic highly values the *timeliness* and *relevance* of the content.

When the Net user depends on up-to-date content, locating timely information constitutes both the goal and the expectation of the user. Financial information ranks as the most popular premium content on the Web today. The Internet provides a direct and quick transfer of information from Wall Street and financial analysts to the Net user desktop. Net users use the Web instead of print media to obtain their time-sensitive information (see Figure 3.2). In their report "The Internet Shift Endangers Financial Print Media," Forrester Research estimates that a full 67% of online financial seekers log on for stock quotes, while only 52% of those who refer to print media are searching for quotes (MacKenzie, 1998).

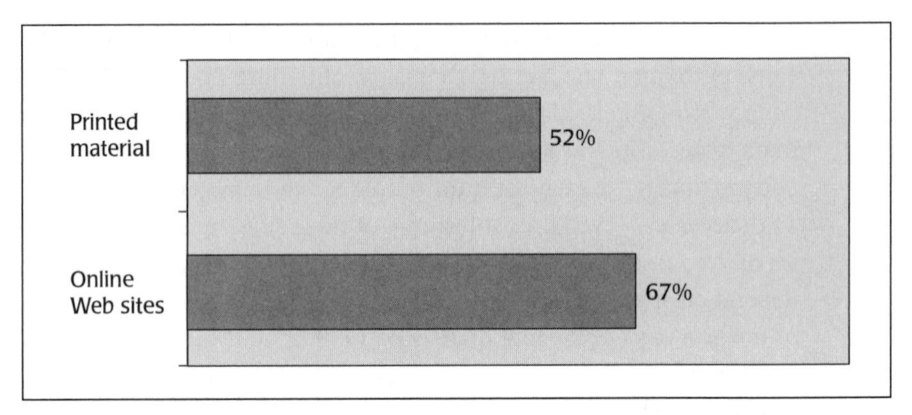

Figure 3.2 Comparison of Use of Printed Materials and Online Web Sites for Stock Quotes.
Source: Copyright © 2000, Forrester Research, Inc. Used with permission.

Home investors, who will manage almost $700 billion in investments online by 2002, according to Jupiter Communications, depend on the timeliness of information and services such as stock lookups and charting (Sterling, 1999). Financial vortals such as CNNfn depend on high traffic volumes by investors checking in on the latest headlines and market fluctuations. For analysis and commentary, some vortals syndicate content from content providers such as TheStreet.com.

CNN limits the value exchange of its online destinations to timeliness for headlines. CNNfn offers extensive facts on investment marketing without analysis and commentary. To obtain in-depth coverage of a trend, CNNfn expects Net users to tune in to the regularly scheduled programs on its cable channel.

Relevance in Content Value Exchange

Content value exchange for vertical portals also depends on the relevance of the content to the Net user. Many sports fans are addicted to statistics and analysis of players and games. Popular sports vortal *www.espn.go.com,* which is owned by the GO Network, both produces its own content and aggregates content from its cable network affiliate. ESPN's SportsCenter provides sports fans with a single destination for just about all the sports-related content available on TV or the Web: the latest news, commentary, statistics, features, player profiles, and video and audio clips. There's even a Java applet (called a SportTracker) that streams scores of live baseball and basketball games to the Net user's screen every 15 seconds to 2 minutes. The Sporting News, a competitor, offers less in value exchange. Its site focuses almost entirely on analysis and relies on syndicated sports news from Associated Press for the latest headlines.

Information without Usability

Content aggregators (networks and portals) run the risk of getting bogged down in flashy presentations at the expense of usability. The Net user's goal in content value exchange is to locate timely and relevant information. Some publishers lose sight of that goal in an attempt to provide an attractive presentation. Focusing too much on presentation and graphics undermines the site's utility. Navigation tools, consideration for bandwidth, and other factors play a key role in making the consumption of content palatable.

MSNBC, a joint venture between Microsoft and NBC, demonstrates one of the risks that content Web sites encounter when "improving" user experience. MSNBC ignores bandwidth concerns, applying extensive graphics and plug-ins to its home page. As a result, Net users wait more than 15 seconds on a 28.8 modem to download headlines from the home page. In addition, Net users have to download client software (an ActiveX control or a Java applet) to enable the interactive menu.

If you have already downloaded or obtained the components through another software program, the experience at the MSNBC site is quite enjoyable. You navigate easily through the site using the drop-down menus that pop up when you move the mouse pointer over a section. But many Net users with lower-bandwidth connections cannot tolerate the download barrier to entry for MSNBC. MSNBC sacrifices easy navigation for presentation.

By contrast, the presentation for the online destination for news competitor CNN looks positively drab. CNN focuses more on satisfying user goals and expectations around the timeliness of content rather than its presentation. The low-bandwidth-friendly Web site doesn't offer large-scale graphics or require client downloads of controls or plug-ins. As a result, the user is more easily able to access information and vivid graphics via text links. CNN targets the same demographic online as with its cable channel—the affluent, white-collar professional. CNN knows that its target demographic focuses on obtaining credible information quickly. As a result, the site's value exchange offers easy navigation, mirror sites in different languages with a regional focus for international users, and a host of other vertically oriented sites, such as the popular financial site CNNfn.

That's not to say that presentation plays no role in value exchange. But publishers can satisfy user expectations around presentation by making it a value-added *service* rather than a core element. Web publishers have a number of techniques at their disposal to build out presentation. Tools like Macromedia's Flash help publishers integrate animation on a Web page without requiring significant bandwidth. The later browsers (Internet Explorer 5.0, Netscape Navigator 5.0, and later) fully support dynamic HTML (DHTML), which lets Web developers build animation effects for text without tools, plug-ins, or servers. Commonly used plug-ins, such as Shockwave, that provide animation on the Web are even bundled as part of the browser.

The trick is to use these presentation techniques to enhance the content experience while not creating a barrier to its utility. MSNBC missteps by making the ActiveX navigation tool part of effective site navigation. While MSNBC does not list the tool as a requirement, the site's hodgepodge of content seems to demand it.

But content value exchange depends first on the provision of information and secondarily on the presentation and format of the content. Until broadband or wireless access is widely and conveniently available, there are and will be far more users who access the Internet via slow dial-up modems than fast broadband connections. Likewise, there is still a population of Net users who have Web browsers that can't handle server-side scripting like JavaScript. Publishers can certainly add flashy graphics and nifty applets to the content value exchange, as long as they do not impede the primary goal. ESPN does a fine job of providing multiple ways to access content for sports fans. For fans with lower bandwidths, the site provides text-based scorecards of live games, with scores updated every couple of minutes. Fans who have access to faster connections can download the site's SportTracker Java applet, which streams scores, players, positions, and statistics in real time during basketball and baseball games. Like ESPN.go.com, successful content sites focus first on getting the information to the Web visitor and second on format and display.

Entertainment Value Exchange

Entertainment Web sites depend on innovation and presentation as hooks for driving Net user traffic. The value exchange around digital music downloads or some other entertainment-based content addresses different user goals and expectations from those around financial content.

Online entertainment mixes the best of television and the Web. Consider why television is so popular today. It offers diverse programming and the ability to passively absorb free entertainment. Sitcoms, drama series, and TV newsmagazines strive to keep us amused, engaged, and (somewhat) informed. But with television, users have no control over the programming and only a limited way to navigate content (by channel rather than by topic).

Let's say that you're a bird hobbyist in search of video footage of African hornbills. Obtaining that footage through television requires diligent monitoring of the programming schedule in search of a nature show that covers the region or the topic. The better answer is to search the Web for a streaming media presentation on the African hornbill or simply head to the National Geographic online site (*www.nationalgeographic.com*) and check for video clips. Online entertainment value exchange capitalizes on the Internet's ability to deliver exactly what the Net user wants.

The Role of Bandwidth

The user's goals and expectations in an entertainment value exchange revolve around *innovation* and *relevance*. The Net user engages in an entertainment value exchange to pursue a favorite interest, such as sports or music. With amusement as the primary goal, the user's expectations include rich forms of presentation and display. That level of presentation is best experienced through the higher bandwidth provided by cable-based or digital subscriber line (DSL) connections.

Broadband access to the Internet, which analysts predict will be widely used by 2003, is the foundation for a truly rich and interactive platform (Reamer, 1999). Broadband is the technology that can transform content from the low-integrity media, such as text and tinny-quality sound files, to high-quality, high-resolution media, such as MP3 files and streaming video. Broadband media, such as streaming audio and video (a software solution that allows Net users to listen to or to view a digitally encoded audio or video file on a remote server without requiring the file to be downloaded to their local drive), provide the best user experience when accessed over cable-based lines or DSL. These technologies mean faster connections than dial-up access by an order of magnitude.

Net users have several choices when looking at broadband connections: cable modems, digital subscriber lines, or satellite connections (see Figure 3.3). (Wireless access is still in its infancy and will not be a truly viable option for several years.) Technically, cable modems are faster than DSL but in implementation they tend to be slower because they use shared bandwidth, which means capacity may be shared by hundreds of other users at a time. Access speed is reduced with each party added to the cable modem. In addition, installation of cable modems can be expensive and require replacement of existing phone wiring.

Digital subscriber lines offer a dedicated line that can prove faster than cable modems in implementation. There's also no rewiring of phone lines. DSL allows existing phone lines to be upgraded to handle high-speed Internet traffic and phone service simultaneously. The drawback to DSL is its inflexibility

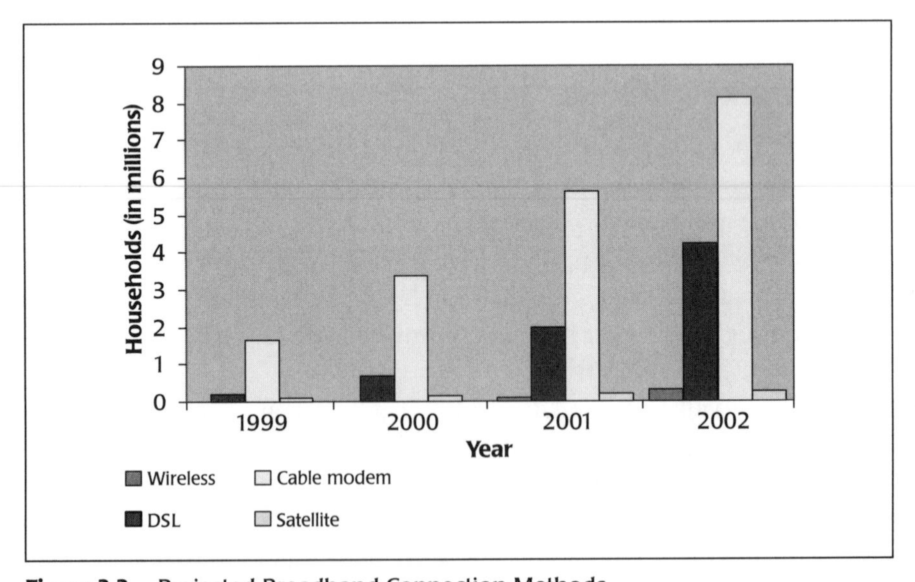

Figure 3.3 Projected Broadband Connection Methods.
Source: From Reamer, Scott, "The Internet Capitalist" newsletter, SG Cowen Securities Corporation. Used for educational purposes only, by permission of and copyright © 1999, SG Cowen.

in location. To upgrade your phone line, you need to have close access to a telephone switch. As a result, DSL access is difficult to obtain outside of major city and suburb limits.

Satellite connections are a third broadband access alternative. To use a satellite connection, you need a PC satellite dish, an unblocked view of the southern sky, and a regular phone line for uploads via analog modem. (Satellite connections are a natural fit for computer users in remote or rural areas.) The only viable satellite solution is offered by Hughes Network Systems' DirecPC, which uploads URLs and address data through a traditional dial-up modem and downloads Web data through the satellite at high speed.

A Net user with broadband access to the Internet can load 30 Web-page frames per second. High bandwidth connections offer more than better access for Net users. They open the door to more creativity and integration between different types of media.

Broadcasting and Integrated Entertainment

Broadcasting on the Internet distributes live television and radio events in streaming audio and video. Television and radio broadcasts give Net users new control over when and where they want to consume the content. The Net user can use a streaming audio player such as RealAudio to listen to a live baseball game while at work. Or, the user can access the archived broadcast of the game while at home. The key to this value exchange lies in providing the Net user the most flexibility in choice and consumption of the broadband content.

Broadcast.com (currently Yahoo! Broadcast),[5] founded in 1995 by Mark Cuban, caters to both Net users and businesses. As a Net user entertainment vortal, Yahoo! Broadcast offers a variety of channels for sports, news, or other categories of content. Each channel of content is threaded with interactive and rich experiences, filled with streaming audio (radio, music, newscasts) and video (concerts, conferences, sporting events).

By 1999, Yahoo! Broadcast provided daily live broadcasts and 65,000 hours of on-demand audio and video programming. It logs about 1 million hits a day. Some of the programming, such as distance learning, focuses on personal improvement. Yahoo! Broadcast delivers a suite of training courses for PBS, The Business Channel (a wholly owned subsidiary of the Public

[5] Broadcast.com was acquired by Yahoo! in 1999.

Broadcasting Service) via the Internet for federal and commercial users' desktops. The first course, offered in March 1999, was (appropriately) "Internet Commerce," presented by the Massachusetts Institute of Technology's Center for Advanced Educational Services. Yahoo! Broadcast owns the exclusive rights to Webcast programming (the broadcast of television or radio programming over the Internet in the form of streaming media) from thousands of radio and TV stations.

Broadcasting's value exchange consists of the ability to offer integrated television, music, video, and (eventually) telephony over the Internet. Yahoo! Broadcast found that Net user adoption of high-bandwidth connections did not pose as significant a problem as expected. According to the Web publisher in 1999, about 99% of office workers have PCs on their desktops, and a full 54% are connected on T1 or faster lines. As part of its value exchange, Yahoo! Broadcast provides suggested connection bandwidths for each audio or video file—and if required, computer hardware requirements.

As part of the value exchange, Yahoo! Broadcast provides diverse ways of getting to the desired content. Channels include text, live broadcasts, and archived video and audio programming. A Net user can download an audio book by Nadine Gordimer by signing up as a Yahoo! Broadcast member (see Figure 3.4). Once the Net user registers, he or she can either download one of Gordimer's books or purchase the print copy by clicking on an Amazon.com link. For the sports fan, there's streaming audio for professional baseball games, horse racing, motocross, and a variety of other events. Yahoo! Broadcast provides different layers of value depending on the content selected by the Net user.

To maintain high performance under high traffic loads, Yahoo! Broadcast reports that the site uses a huge network for broadcasting, including 22 satellite receiving dishes, more than 1000 multimedia streaming servers, and support for 650 simultaneous live events.

For example, if the Net user wants an audio book, Yahoo! Broadcast provides links to author interviews. Net visitors tuning into a company's quarterly earnings call can link to the company news, stock quotes, and analysts' comments about the company. In addition to broadcast events, the vortal hosts 2,400 full length CDs, 350 live concerts, and 5,000 hours of on-demand video.

Figure 3.4 Listening to an Audio Book at Yahoo! Broadcast.
Source: Reproduced with permission of Yahoo! Inc. © 1999 by Yahoo! Inc. YAHOO! and the YAHOO! logo are trademarks of Yahoo! Inc.

The Music Revolution

Online distribution of music on the Web is a hot topic with Net users, artists, and record labels. Despite its hype and promise, it's almost impossible to predict what mainstream value exchange around online music will look like. Even Michael Robertson, the CEO of MP3.com and one of the most vocal proponents of online music, expressed confusion in a November 1999 interview with *Business 2.0* magazine (Hellweig 1999). "What we have now is a baby," Robertson explains, "and looking at a baby and trying to predict what it's going to look like when it grows up is an exercise in futility." What is readily apparent is the Net user demand for flexible distribution of music on the Web and the big music labels' equally strong desire to prevent the unauthorized delivery of music files to Net users.

Packaging Audio Files for Internet Distribution

To package music for consumption on the Net, a music publisher uses a music-editing tool to compress the track into a file format. The Net user then downloads the file to his or her machine and uses an online music player to listen to the music. (You can download one of them for free—my favorite is at *www.musicmatch.com.*) Alternatively, the Net user can load the file online in the online music player, and the publisher "streams" the compressed audio from a media server to the player

One file format, MP3, has emerged as a clear leader in the file format standards war and is supported by most streaming media players.[6] It's difficult to get exact numbers for MP3 file downloads on the Web today (estimates hover around 2 million) because of the different methods for download and also some artists' efforts to get on a download site's "hot list" (the most frequently downloaded songs). According to a spokeswoman from the MP3.com download center Riffage.com, an unknown number of bands run scripts that download their songs over and over (Ledbetter, 1999). As a result, publishers can't accurately determine how much traffic is from actual fans or self-promoting bands. Michael Robertson gets around this distortion by using sales generated from CDs to measure the popularity of the band's online songs. According to Robertson, MP3.com compares the number of downloads to the number of CD sales. Then the music portal ranks bands using that conversion rate as one of their evaluation statistics.

According to Robertson, by 1999 MP3.com had over 23,000 artists and is adding over 200 artists a day. Robertson believes that "at the end of the day it's really all about a new way of distributing." The primary value exchange is a new way that users can access and transport the music.

The MP3 file format (and whatever higher-quality format replaces it in the future) certainly does help bridge the gap from computer to CD player to portal player. Part of the MP3's popularity rests in its open-source model and inexpensive, nonexclusive licensing technology. A variety of vendors have already leveraged the open model to build players for MP3 files.

Diamond built one of the first digital music players, called the "Rio," that allows Net users to download songs in MP3 format from their computer to

[6]Before MP3 emerged as the primary music format, individual audio player manufacturers characteristically supported only their own proprietary file formats.

the player. Users can then play the songs saved to the small, digital MP3 music device anywhere they go—without having to connect to the Internet.

Diamond's Rio digital music player (which survived a major record label's legal attempt to squash its distribution) offers Net users the ability to take their Internet music tracks on the road. Some vendors are also looking to port MP3 player functionality to the popular 3Com PalmPilot platform. Net users could download tracks from the Internet to their PalmPilots, building music into the handheld device platform.

The flexibility in distribution that online music file formats introduce also raises significant issues around rights management (as the major record labels continue to lament). While file formats such as MP3 simplify the distribution of music from online to offline channels, they also facilitate the ability to redistribute music files on the Internet without regard to licensing or copyrights.

Rights Management

The primary obstacle to mainstream music distribution on the Internet is the lack of support from the major labels (BMG, EMU, Sony Music, Universal Music, and Warner Music). The labels hold the rights to most commercial music and have refused to allow Internet-based distribution in the fear that music files will be freely traded between Net users. Millions of Net users have downloaded music from the Web. New media players allow Net users to "rip" (to copy) CD tracks to MP3 format (newsgroups are popular places to swap files).

In most cases, moving audio files between media is more for convenience rather than an act of conscious piracy. Nevertheless, to prevent the unauthorized distribution of audio files, the five major labels are simply withholding their vast archives of music from distribution on the Internet. In addition, most labels also legally bind artists from distributing music in MP3 format. Without a secure way to distribute music online, the big labels refuse to engage in online music value exchange at all.

Without the support of the top five record labels, Net users will rarely be able to get their favorite songs online. The vast majority of songs available today are from lesser-known bands. While Net users can find name-brand artists, the tracks are usually pirated and posted on continually shifting newsgroups. Net users have to know where and when to get the pirated content since the download points shift frequently. Mainstream value exchange for online music depends on record labels joining the music revolution and coming up with a solution for the rights management problem.

There are two options for *digital rights management* today:

1. Secure distribution
2. Watermarking

Secure distribution requires that consumers have software installed on their machines that reviews their right to access the content before opening an encrypted file. For example, the file may not be opened until the consumer agrees to pay for it and/or signs a usage agreement. This kind of management focuses on *controlling access* to premium content.

While effective, secure distribution suffers from its dependency on custom software on the consumer's computer. If the consumer doesn't have the custom software or has the wrong version, the burden falls on the consumer to find and install the software. This level of requirement might be acceptable for business-to-business exchange of high-value information, where the corporate environments can require the custom setup as part of the standard computer configuration for the company. But for business-to-consumer transactions, it's impractical to assume that every consumer has or wants to have the rights management software installed.

Digital watermarking focuses on identifying and discouraging improper usage of the content. A digital watermark is a set of data that is embedded in an audio or video file to identify the origin or ownership of the file. Image publishers use visible watermarks (like company logos displayed prominently on the image) or invisible watermarks (called *steganographs*). Digital music watermarks are placed at very low frequencies, barely audible to the human ear. The consumer doesn't need any special rights management software to access and use the watermarked file. In fact, nothing is preventing the consumer from redistributing the file without authorization.

Watermarking depends on two factors to limit unauthorized distribution. First, owner identification may discourage illegal distribution because the consumer is more aware of the copyright and also because the quality of the file is somewhat degraded. Second, digital watermarking solutions often offer a software program that "crawls" over Web sites on the Internet, identifying watermarks and verifying improper usage. Web publishers use that information to send cease-and-desist letters for unauthorized usage or even to take offenders to court.

While digital image providers have paved the way in watermarking, many music publishers are still nervous about using any rights management strategy that hasn't been accepted as a standard by the major music labels. However, a number of initiatives are underway to standardize the secure distribution of music. Some efforts, such as IBM's Madison Project, provide lip service to the recording industry's claim to seriously examine online distribution as an option.[7]

A standard that has picked up momentum among the major music labels is the Secure Digital Music Initiative (SDMI). SDMI presents a set of security guidelines for selling music over the Internet, mainly focusing on recoding digital compression devices needed to download music. The SDMI initiative follows the model of client-side rights management, focusing on restricting distribution (such as preventing unauthorized downloads) as opposed to tracking redistribution through watermarks. For example, the initiative proposes to block portable MP3 players from accepting illegally created music files. But the online music distribution issue is far from resolved. While the major labels support SDMI, many independent publishers and artists advocate watermarking solutions or no tracking at all for music redistribution. The real question is whether the big music labels will continue to refuse to release their libraries until there is agreement on the issue.

Summary

Killer content and services vary by category of value exchange offered on a Web site. Promotional exchange is the only noncommercial type of value exchange—that is, it cannot be tapped for revenue since its purpose is pure advertisement of a company or product. A commerce exchange benefits the Net user by providing product or content in exchange for the user's payment. Enhanced services include community forums and responsive customer services. Content value exchange revolves around the flow of information between Net

[7]The Madison project is testing digital downloads with a limited audience over a six-month period, after which the major labels have promised to review the results.

user and Net publisher. An entertainment value exchange focuses on the presentation and innovativeness of the content. Entertainment value exchange includes the rapidly expanding field of online music distribution, which is in search of a business model that satisfies both Net consumers and music labels.

The varieties of value exchange drive the next generation of commerce on the Web, which blurs the line between content and commerce. The evolution of commerce and distribution models lets Web publishers diversify revenue streams for sustainable revenue.

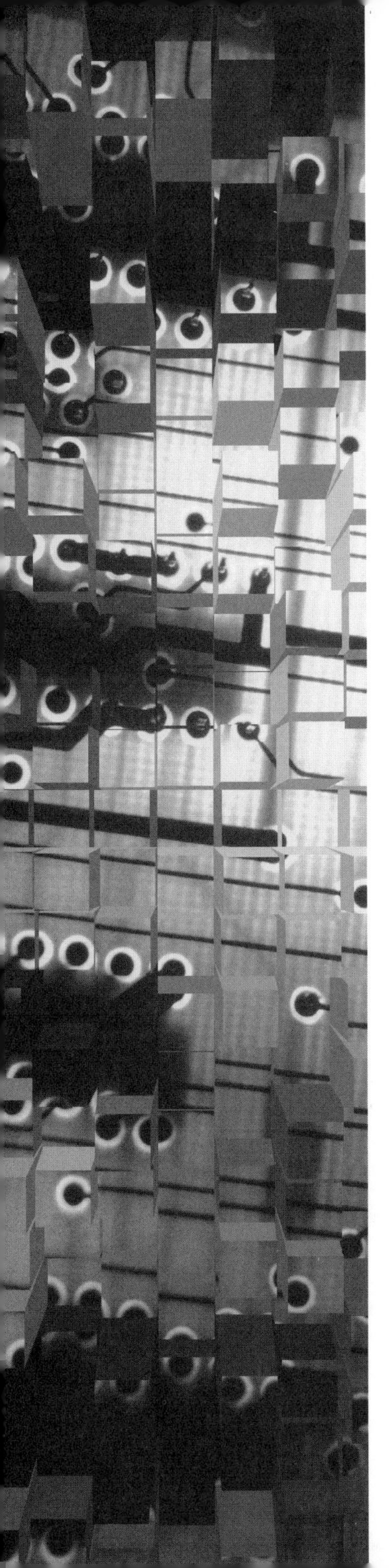

CHAPTER FOUR

Diversification of Revenue Streams

Without monetizing value exchange, a Web site can offer the best content services on the Web and still not generate revenue. To succeed in the Internet economy, publishers need business models that create multiple revenue streams from the value exchange on their sites. High-flyers of the Internet world such as Amazon and Yahoo! and new businesses such as TheStreet.com and Salon demonstrate how a Web-based business can branch out from its core competencies of books and financial advice to earn dollars from content, commerce, and services. This combination of revenue streams lets publishers make money from their value exchange.

Cutting a profit on the Web also requires the right balance between revenue and the costs of operation, acquisitions, and other factors. Monetizing value exchange is about maximizing the inflow of *revenue* for an Internet business.

The depth of the value exchange directly impacts the success of the business model. Single-dimension value propositions, such as rock-bottom prices, can't survive in a Net economy. The first three chapters illustrated how enhanced services increase the appeal of a site's value exchange for Net users. The quality of the value exchange translates to a competitive advantage over less diversified competitors. Sites such as Amazon and Cisco gained their competitive advantages by building out services around the commerce exchange. Net vendors with a narrow view in value exchange, such as Egghead.com and Priceline.com, have failed because the single-minded focus on one dimension ends up creating a low barrier to entry for competitors.

Three factors stand out as performance indicators for success in an Internet business: revenue, membership (the number of Net visitors who have registered at the site), and traffic. Which success factor is stressed depends on the company's objectives. Amazon measures its success by customer loyalty rather than profit. It considers its base of repeat customers a long-term investment for continuing revenues in the future. Publishers like Yahoo! who depend on traffic to drive advertising revenue measure success by the number of page hits, or people who click on a particular advertisement. Regardless of preferred performance indicator, Web publishers should implement internal or external systems that help track progress against the company goals. Content management and Web site servers provide the in-house tools to track rates like membership acquisitions. In addition, many out-source services handle revenue streams, such as affiliate link programs[1] and tracking, as part of the service fee.

Content publishers, such as Salon.com and TheStreet.com, use integrity and relevance in content as differentiating factors against other content providers and vertical portals. (The barriers of entry for content competitors are as high as the quality of the value exchange.) This chapter takes the notion of a quality value exchange a step further by translating how the exchange maps to revenue dollars.

An Evolution Underway

The growing emphasis on Net user goals and expectations has necessitated an evolution in the ways that commerce and content distribution happens on the Internet. That evolution has affected retail commerce exchange (the "traditional" form of e-commerce) by extending the definition of a merchant. In addition to commerce exchange between Web retailer and Net consumer, commerce transactions also occur between Net consumers and other con-

[1]Content publishers host affiliate links for retailers in return for a percentage (between 2% and 10%) of any purchases generated by customers who click the link. Chapter 5 has more details on affiliate link programs.

sumers. Variations, such as auction pricing and consumer merchandizing, represent new ways to monetize commerce exchange.

Traditional e-commerce businesses use the Internet as an online catalog. These businesses collect orders on the Internet and fulfill the orders using an established supply–distribution chain. This discussion focuses on new ways of monetizing value exchange and does not cover the established retail e-commerce model.[2] To remain competitive Internet businesses have developed new e-commerce models.

Content sites have struggled to find revenue on the Internet. With a few notable exceptions,[3] portals and content providers alike have had difficulty in building sustainable revenue from the millions of online users. Maintaining the timeliness, relevance, credibility, and utility in content value exchange takes resources and capital. Publishers that produce their own content face the most risk. Sports broadcaster Quokka.com shoulders high content production costs that carry the company into considerable red ink on the balance sheet. In 1998, the Internet company lost a total of $9.5 million, compared to $8.6 million in revenues.

Founded in 1995 by CEO Alan Ramadam and Chairman Dick Williams, Quokka broadcasts second-tier sporting events such as auto racing, sailing, and mountain climbing.[4] Quokka uses broadband media, such as streaming video and audio, to provide "total sports immersion" for long-duration events. For example, one of Quokka's adventure features covered the Marathon des Sables, a grueling seven-day race through 220 kilometers of the Moroccan Sahara Desert. The combination of video, audio, and text is expensive to produce (Quokka doesn't provide details on actual costs per event program).

The company's primary revenue comes from selling sponsorships to corporations and sports organizations and has fallen short (by $1.1 million in 1998) in covering operating costs. To balance out the difference, Quokka is adding new revenue streams, such as syndication (licensed content to Yahoo!

---

[2]For a thorough description of building a retail infrastructure on the Web, see *E-business: A Roadmap for Success* (Kalakota et al., 1999), or *Net Success* (Haylock et al., 1999).

[3]Profitable Internet businesses such as Yahoo! built their successes over time on the strength of their content value exchange.

[4]Quokka also entered into a joint venture with NBC for exclusive rights to Webcast Olympic events through 2004.

and other portals), and is exploring either a subscription or pay-per-event model. Other publishers who offer content value exchange, such as Webzine Salon.com and financial commentary site TheStreet.com, share Quokka's focus on a hybrid business model (see also "Monetizing Content and Entertainment Exchange" later in this chapter).

Content value exchange calls for new flexibility in "merchandizing" digital content. Content publishers are learning how to merchandize their premium content to earn revenue from their variations of value exchange. The merchandizing of content for Net users means subscription-based access to journalistic scoops, as found at TheStreet.com, or pay-per-item purchases of archived articles in *The New York Times on the Web*. Business customers of content include portals that aggregate analysis from different content providers for corporate intranets and distribution lists. Content publishers use syndication technologies to display their content automatically in business customer Web sites. New secure ways to distribute information via Web and e-mail allow content providers to leverage convenient channels without fear of copyright infringement or unauthorized access. These different content merchandizing strategies open up new avenues to monetize value exchange. Any value exchange, with the exception of promotional exchange,[5] supports one or more revenue streams from content.

Monetizing Commerce Exchange

Net users are driving the evolution of new forms of Web payment, packaging, and distribution. Value factors like price and community are key in the new ways of doing business on the Internet.

In "traditional" e-commerce exchange, established by online retail sites, the retailer sets a fixed price for an inventoried product. The Net user uses certain value factors, such as price, brand credibility, or availability of other services, to make his or her purchase decision. Traditional retail exchange appeals to Net users whose value factors consist of credibility of a commercial brand and the security of dealing with an automated order fulfillment process.

---

[5]Promotional value exchange is limited to corporate Web sites or online sites dedicated to extending an offline (radio or television) advertising campaign.

Elements of Consumer-Driven Commerce Exchange

Auction pricing: Negotiable pricing determined by the buyer's and seller's efforts to consummate a mutually acceptable exchange.

Consumer merchandizing: Net users who use hosting services and online consignment services to sell used retail and/or home-manufactured inventory.

Escrow: Funds or goods delivered by one person to another entity, which on the fulfillment of a certain condition, is delivered to a third party. For an online auction, funds can be held "in escrow" until the buyer gets and approves the merchandise, upon which the funds are released to the seller.

Merchandizing content: The packaging of content (text, images, and rich media) into a product that can be bought and sold by Net users and businesses.

The success of a new model of commerce exchange points to the growing number of Net users who emphasize different factors in the commerce value exchange. Auctions and consumer merchants provide highly diverse inventories of used goods, as well as low prices and a personal touch in the transaction. These new ways of monetizing value exchange emphasize *flexibility* in fulfilling the Net user's purchase goals. In an auction, the Net user has the power to explicitly declare his or her value of an item through a bid. If the seller approves the bid, the Net buyer gets to fulfill a value exchange shaped to his or her goal and expectations. Consumer-driven businesses through auctions and hosting services also appeal to Net users who emphasize *cost* and *uniqueness* as critical value factors.

As discussed in chapter 3, the Net user's goal in a commerce exchange is to obtain a product at a certain price. Negotiated pricing and consumer merchandizing (Net users selling goods to other users) offer an alternative for big online retail sites to fulfill those goals.

Negotiated Pricing by Auction

A commerce exchange conducted by auction offers an eminently practical business model. Auctions operate on pure supply and demand. The seller offers an existing item in inventory at a price to stimulate demand. Buyers bid on the item according to the level of value that they subscribe to the commodity. The item sells when the seller offers the right value exchange in the transaction for the buyer. Many Net users prefer auctions because of the low prices on used goods and the personable nature of a Net-user-to-Net-user transaction. As one auction user put it, auctions are like "bed and breakfasts in the hotel industry," providing a touch of personal service to what is otherwise a straight anonymous transaction.

Auction sites are successful when maintaining a certain density of buyers and sellers. There are niche auction sites that focus on selling a single category (such as sports paraphernalia), portal auction sites (such as Yahoo! Auctions), and free auctions, which earn revenues solely from advertising—and then there's eBay. The first Internet auction site, eBay is indisputably the leader. The San Jose–based company provides more than 1,600 categories of items to nearly 4 million registered buyers and sellers. According to eBay, more than 33 million items were offered up through auction in 1998, with over 50% resulting in a sale.[6]

eBay pioneered auctions as a business model on the Web. Pierre Omidyar, eBay's founder and chairman, launched the online service (then called Auction Web) on Labor Day weekend in 1995. In 1996, the company began charging sellers for auction services (the source of revenue for most auction sites) and launched the popular Feedback Forum, which allows buyers and sellers to rate each other. Auction Web changed its name to eBay in 1997 and, through aggressive advertising campaigns, reached a daily rate of 800,000 auctions by the middle of that year.

Depending on your opinion, auction sites provide the world's largest flea market or the most complete showroom for new and used goods. Regardless of which camp you fall into, the fact remains that an auction environment opens up a huge range of inventory in commerce exchange. Buyers have a greater chance to find what they are looking for at the right price. The seller can conduct business without the overhead of Web site maintenance and commerce system infrastructure. In addition, auction users have access to a community of buyers and sellers, a key element in the value exchange. As one

---

[6]eBay's technical infrastructure has had problems keeping pace with demand. The site suffered several significant outages in 1999 as eBay struggled to expand its transaction-handling infrastructure.

auctioneer points out, since many sellers buy inventory through other auctions, relationships are formed around the commodity on auction. That sense of community also stimulates repeat purchases. As one bookseller says, "I have lots of repeat customers. I've made so many cyberfriends. And, fortunately for me . . . they like books."

Auctions have opened up a new segment of merchants in online commerce transactions: home-based businesses with informal supply networks (e.g., thrift stores, garage sales, and other auctions). Laurie Clark of Oceanside, California, seems a bit puzzled when you ask her if she considers her auction-based business on the cutting edge of the Internet economy. "Well," the soft-spoken 46-year-old says, "not really. It's so easy, and it requires no capital investment. You can hang a shingle out and be in business in 24 hours." Laurie sells all her inventory of 70 to 80 antique collectables and figurines at online auctions. She works primarily through eBay, where she says the quality of buyer is better (less prone to ducking out on a purchase), and she gets more bids on her inventory. Laurie posts less than 5% of her auctions at Yahoo! Auctions, Amazon Auctions, and Auction Universe, since, in her experience, buyers at those auction sites tend to be less reliable.

Laurie's been selling through online auctions from her home for 2 years, after 18 years in the insurance business. Like most auction sellers, Laurie uses her home-based business to supplement another income. Ask Laurie what she values in the auction selling model, and she'll tell you how she can sell the inventory stored in her garage to hundreds of customers any time of the day. That's with an old 333 Mhz PC computer on a dial-up modem connection (L. Clark, phone interview, 8 Oct. 1999).

Inventory comes from an informal supply network, such as thrift sales, garage sales, antique stores, and other online auctions. Laurie points out that you don't have to go very far to build an inventory. "Just look around your desk," she says, "and you'll see two or three things you can sell in an auction."

Fulfilling Payment in an Auction Exchange

However, flexibility in inventory and order fulfillment introduces the risk of unscrupulous buyers and sellers failing to fulfill their end of the commerce value exchange. At online retail stores, Net users generally pay by credit card when making a purchase. The Web retailer provides several explicit and implicit reassurances during the transaction, such as the reputation of the merchant's brand, an online customer service center, or an e-mailed receipt upon order placement. The security of the transaction is part of the retail value exchange.

Auction sites don't offer the same assurances. eBay very clearly states on its site that the company is in no way accountable for the quality of the merchandise sold on the site. The trust element in auctions is most evident in the merchandise fulfillment process. When a seller accepts a bid, the seller and buyer coordinate how to handle payment and shipping for the item. Generally, the seller will not ship the merchandise until receiving payment. (Most auction sites balk at offering a credit card processing service because of the high risk of fraud and frequent chargebacks to credit cards.) Sometimes this trust relationship breaks down. A seller may fail to send the purchased item after receiving payment. A buyer may agree to send payment and never follow through. Or a seller can accept payment and ship a lesser quality or different item altogether.

To shore up trust in the auction commerce exchange, most auction sites offer online escrow services. These services provide, in the words of one escrow user, "a safe way to buy without getting burned." i-Escrow, founded in 1997, is the leading online escrow service. The California-based service handles escrow for a number of large auction sites (see Figure 4.1), including eBay and Amazon Auctions. A seller in the online auction specifies whether he or she accepts escrow payment when setting up the auction. When a seller and a buyer agree to a price, the buyer goes to the i-Escrow Web site to submit payment (usually using a credit card). i-Escrow then holds the buyer's money in a non-interest-bearing trust account, and lets the seller know that good funds are being held in trust. The seller ships the merchandise directly to the buyer

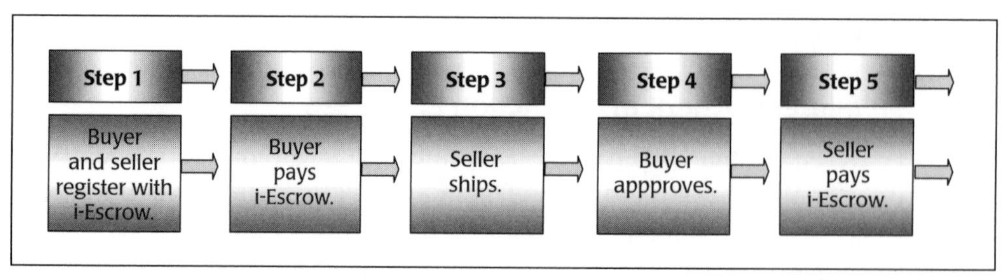

Figure 4.1 Payment through Online Escrow.

for inspection. (i-Escrow simply provides an extension of the negotiation process and never takes possession of the merchandise.)

After the buyer receives, inspects, and approves the merchandise within the agreed inspection period, i-Escrow releases the funds to the seller. In the case of a return, the buyer is refunded only after the seller's receipt, inspection, and approval of the returned merchandise. i-Escrow makes its money by taking a percentage of the transaction, which sellers usually pass on to the buyer as part of the merchandise price.

The buyer can pay with a credit card as opposed to check or cash-on-delivery and has a clearly defined approval process for the merchandise. The seller benefits from the added security for large ticket goods. In addition, a service like i-Escrow gives small home-based businesses the ability to use credit cards online. As Laurie Clark puts it, online escrow services make the purchase experience function the same for merchandise sold from the family room or a 10,000-foot showroom. Laurie uses i-Escrow for her auction sales over $25 and passes the cost of the escrow service on to customers. After Laurie signed up for i-Escrow's services, she saw sales increase by 30%. "It's insurance, basically," Laurie explains. "Buyers out there don't want to send cash to a stranger."

i-Escrow's payment processing services add trust to the value exchange. Within its first two years of business, the escrow service had earned up to 160,000 registered users. i-Escrow gained its users through word of mouth and partnerships with auction houses. The company plans to launch advertising campaigns to increase Net user awareness of escrow services. According to Sanjay Bajaj, i-Escrow's vice president of business development, widespread adoption has been hindered by the lack of customer awareness of escrow services in general. Many Net users have never used an escrow service. The most common use of escrow involves down payments in a real estate transaction. Even then, only 10 out of the 50 states require that real estate funds be held in escrow by law. As Bajaj ruefully points out, "In the remaining 40 states, the word escrow doesn't come into their lives enough." As a result, only a limited number of Net users are familiar with the escrow model in the physical world, much less the virtual world. But, as Bajaj rightly points out, Net-user-to-Net-user sales involving high-value items can benefit from the security and trust that the escrow service brings to the table (S. Bajaj, phone interview, 7 Oct. 1999).

i-Escrow tries to limit the service's exposure in the buyer/seller transaction by providing an automated negotiation engine software, which handles different scenarios of negotiation around shipping and approval. The negotiation engine speeds up the transaction by reconciling new transaction terms instantaneously. For example, if the buyer proposes a transaction term that the seller has not agreed to, the change triggers an e-mail to the seller asking for approval. The company has automated the communication process for over 500 scenarios that can come up during the auction payment process. As a result, i-Escrow claims that 90% of customers surveyed indicated that they were "satisfied" or "very satisfied" with the service.

To build trust in the commerce exchange, i-Escrow pays considerable attention to customer communication. The escrow service's customer support gets more than 1,000 e-mails a day, mostly from first-time auction participants or escrow users. According to Bajaj, i-Escrow customer service responds within 3 to 4 hours of a request. The escrow service also provides close to 24-hour support for customers, from 5 A.M. to 1 A.M., seven days a week.

Consumer Merchandizing

Escrow services also come in handy with another form of Net user–driven value exchange. In 1999, Amazon.com, borrowing heavily from the auction model, pioneered a new type of consumer merchandizing called zShops.

Consumer merchandizing combines different elements of a retail and an auction value exchange. Retail value exchange provides the benefits of a trusted brand and mainstream payment collection through credit cards. Auction value exchanges provide negotiable prices and a wide variety of inventory. Consumer merchandizing combines the security of an established brand with the breadth and flexibility of an online auction inventory. This particular type of commerce exchange uses fixed pricing, as in a retail transaction, but does not necessarily depend on an established business supply and distribution chain.

Amazon.com zShops is one of the first examples of a consumer merchandizing forum on the Web. Launched in the fall of 1999, zShops gives Net user merchants like Carol Hamilton the ability to sell under the umbrella of the respected Amazon brand.[7] Carol runs a zShop called Carol's One Stop Book Shop out of her home in Leslie, Arkansas. She started off selling her "zillions of books" (currently 13,000 stored in the carport) through online book exchange list servers like Bibliofind after she became disabled in 1998.

---

[7]Amazon.com also insures transactions for up to $1,000 at a zShop.

After Amazon acquired Bibliofind, Todd Rohs, from Amazon's Merchant Development group, came across Carol's listings and, according to the Dallas native, "gave me a holler." Since the launch of zShops, Carol's sales have doubled and apparently driven the town of Leslie, Arkansas (population: 400) to hire a part-time postal worker. Carol obtains her inventory the same way auction sellers do—through junk stores, flea markets, library sales, and Goodwill. But Carol prefers zShops to an auction environment because of Amazon's credit card processing service (which half her customers use for payment).

Carol's One Stop Book Shop represents a new type of commerce exchange that combines the personal touch and wide inventory of auctions with the security of a retail transaction. Carol's ability to run a Net-user-to-Net-user business that, according to the online merchant, "make[s] enough to make me happy" shows how Net users' goals and expectations in commerce value exchange evolve into organic business models. A similar evolution is happening around content (C. Hamilton, e-mail interview, 13 Oct. 1999).

Monetizing Content and Entertainment Exchange

Great content does not guarantee a great business model. In fact, many publishers of high-quality content, such as Webzine Salon.com and TheStreet.com, still struggle to turn a profit on content alone. These sites and others are experimenting with new methods of monetizing the value of content. This section explores the different ways that content publishers can distribute and monetize content, including selling access to individual areas or specific pages on a site, content or program downloads, licensing content to other Web sites, and a variety of other models. Early attempts at this type of content commerce did not exactly meet with success.

Early Failures at Content Commerce

Conventional Web wisdom has scoffed at the value of content. In the early days of the Web, analysts and Net user alike declared that Net users would never pay for content because they could simply travel to another Web destination that offered the content for free. Publishers heard such claims and decided that it was more important to not lose traffic for advertising than to force a business model that called for restricted access to premium content. Through the mid-1990s, conventional Web wisdom was right. Very little content and few services on the Web were worth a premium, and there was only a

Revenue Streams for Content Value Exchange

Advertising: Price set by Web traffic volume and demographics.

Pay per item: Price set for access to a single product, or download, on a per item basis. Commonly referred to as "à la carte" products.

Secure distribution: Price set according to business rules embedded in the product file format. Since the pricing and access restrictions are integrated with the product, Net users can redistribute the product through e-mail or other Web sites.

Subscription: Price set according to time-based access to content. Applies to an entire Web site or some subsection of the Web site.

Syndication: Monthly (or other time increment) fee for pushing content to a syndicator Web site.

single form of monetization available for content value exchange: subscriptions.

As a result, early experiments in content commerce failed miserably. Microsoft's editorial magazine, *Slate,* took the print media model of subscriptions to the Web and tried charging subscription-only access to their site. In fact, *Slate* tried charging subscriptions for access twice, once in February 1996 and once in the fall of that same year. Both times, enraged *Slate* readers complained loudly through e-mail or simply switched loyalties to Salon.com, then a fledgling competing editorial publication. *Slate* readership dropped off dramatically, and the magazine has since been slowly and laboriously claiming back readership.[8]

The model languished until the late 1990s when some companies like TheStreet.com began experimenting with mixed revenue streams instead of sticking with a single revenue generator. Syndication, pay-per-item, mini-subscriptions, and other diverse pricing models opened up the market for premium content by providing flexibility in monetizing content value exchange.

[8] *Slate* now offers subscription access to only a subset of their Web site, with most content available for free.

No longer were content publishers stuck with the choice of only subscriptions or free access for content. Even when subscriptions generate revenue it is rare that subscriptions alone will provide a viable business model for content publishers.

Advertising

Most content publishers use advertising, which has its own inherent challenges, as a revenue source. Only top portals that get millions of visitors per day can survive on advertising as a single revenue stream. The vast majority of Web publishers cannot depend on advertising alone to monetize content value. Fewer than 100 ad-supported sites are drawing the type of traffic that sustains advertising as a primary revenue model—with more than 10 million page impressions per year. Although advertising revenue on the Web continues to grow (analysts expect the amount to rise to $10.8 billion by 2002—Charron, 1999), the growth will continue to be centralized in an exclusive tier of sites. Yahoo!, as the top-ranked Internet portal/network, dominates that exclusive tier.

Based in Santa Clara, California, Yahoo! gets more than 32 million visitors each month. The network has more than 3,800 advertisers jostling for space on Yahoo! Web pages, which drives the company revenues. In fact, Yahoo! generates the majority of its current revenue through banner ads.[9] Stanford grad students Filo and Yang developed the Yahoo! search engine and unveiled their Web site in 1994. By the end of that year, Yahoo! had attracted hundreds of thousands of Web surfers. The network started collecting ad revenue in 1995. By 1999, Yahoo!'s 2,700 advertisers and 6,000 merchants combined to offer a revenue stream of $500 million a year. Amazingly, Yahoo!'s profit margin now surpasses most print-based media companies.

Yahoo!'s success demonstrates what advertising can do for the top-ten trafficked Web sites. Advertising as a revenue stream for everyone else looks decidedly less rosy. Ad banners are sold by individual units and tracked by either exposure or click-through rates. Revenue typically comes from the CPM (cost per mille, or thousand page impressions) rate charged by the Web site publisher for ad space.

[9]Even Yahoo! is trying to move away from a single business model by diversifying its revenue stream with e-commerce, promotions, and sponsorships.

Advertising Metrics for Success

Ad inventory: The Web publisher's available slots on the site for advertising.

Click-through rate: The number of times that Net users clicked on the advertisement to find out more. (Assumes that the banner advertisement provides a link.)

Conversion rate: The rate at which the Net user actually completed the action advertised in the banner. Typically, conversion rates describe completed purchases and registrations from a banner advertisement. Used by advertisers to gauge return-on-investment for campaign.

Page impressions: The number of times the page hosting the advertisement is displayed.

Predictably, targeted advertisements displayed by keywords on searches come at a premium. A Forrester Research survey found that untargeted banner ads were going for $5 to $25 dollars, while targeted ads by keyword started at $20 and went as high as $50 per CPM ("Pricing Site Traffic"—Nail, 1999).

In addition, ad inventory faces a strong buyer's market. With a large number of Web publishers competing for advertisers, CPM prices can very quickly drop to the bottom of the range (about $5 per CPM). Most Web site publishers maintain a certain percentage of ad inventories to unload in barter, self-promotion, and fire sale liquidations.

With about three-quarters of ad revenue going to top-ten Web destinations, advertising is best used as a supplemental revenue stream. Advertising can monetize any type of commercial value exchange: commerce, entertainment, and content. But, only sites that get high volumes of traffic (around 8 to 10 million visitors a year) can support a revenue model that consists solely of advertising (see Figure 4.2). Demographics also play a strong role in ad inventory pricing. Technology-aware, affluent professionals strike the best note with potential advertisers, as this demographic tends to be the most prone to online (and offline) purchasing. Finally, it's important to consider bandwidth concerns when displaying advertising. Burdening a Web page with interactive

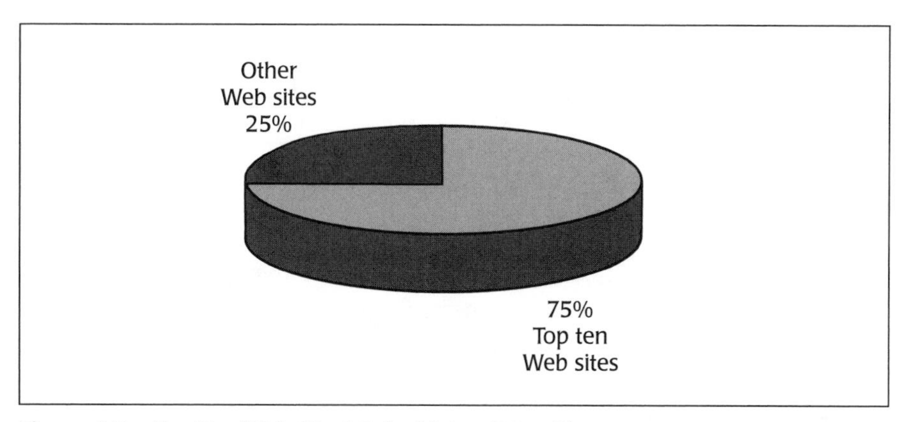

Figure 4.2 Top Ten Web Sites' Cut of Advertising Pie.
Source: From "Ecommerce: Corporate Experience with Websites" online information. Used with permission of eMarketer, New York, NY.

banners risks losing users with low-bandwidth connections and older versions of browsers as the page loads, although sites that expect high-bandwidth connections have greater flexibility with more entertainment-oriented advertising forms.

Advertising by Demographics

Publishers with desirable demographics are best suited to leveraging advertising as a revenue stream. A strong demographic supports a consistent and high CPM rate. By 1999, financial content provider TheStreet.com saw about 50% of its revenue from advertising because of its target demographic of high-income, educated professionals. TheStreet reported that 40% of subscribers had an average income of more than $200,000, and almost 45% had more than $300,000 in assets in their portfolios. In addition, almost one-third of the subscriber base had post-graduate degrees, and almost a quarter of subscribers owned their own businesses. The average subscriber at TheStreet.com logged on three times per day and spent about 22 minutes on the site per visit, which means multiple minutes of exposure for advertisements on the page (Reamer, Aug. 1999). A spot in the TheStreet.com advertising goes for around $55 per CPM, which is in the top range of the online advertising world. Advertisers include luxury brands such as Merrill Lynch, Mercedes, Reuters, and Volvo.

Rich Media Advertising

Rich media advertising increases the percentage of click-throughs for advertisers by integrating interactive elements and animation into a banner to catch site visitors' interest. Rich media advertising includes banners and applets that depend on audio and video, billboards, and/or scrolling text to market a product. Net consumers looking for entertainment value exchange (such as that offered on Yahoo! Broadcast) often come prepared for rich media presentation with fast connections and powerful computers.

Rich media advertising consultants like Excite@Home's Enliven Business Unit (formerly Narrative Communications Corp.) build interactive advertising banners, like the Hard Rock Cafe Rock Trivia Quiz.

The quiz offers the Net user the chance to interact with the Hard Rock Cafe without leaving the site on which the banner advertisement is viewed. Click the Print link, and the banner ad opens another browser window with a coupon that you can then print out and use at one of the restaurants. The animated banners pique the Net user's interest by offering a series of trivia questions about rock 'n' roll history. Net users click a checkbox and wait for the banner to respond with the right answer and a request to continue to the next question.

With its interactive elements and animated presentation, rich media advertisements attract more click-throughs than a static banner. Unfortunately, most Web sites users do not have the bandwidth to support rich media advertisements. *Interstitials,* the popup ads that open when a Web page loads, greatly increase the amount of time a Net user has to wait to view the desired Web pages. A New York–based startup called Unicast developed a Java-based rich media ad called a *superstitial* that allows publishers to use interactive advertisements without incurring the same performance drawbacks as interstitials. Superstitials download to the cache memory of the Net user's browser *after* the browser downloads and renders the actual Web page. The caching of the advertisements lets the user see his or her page content quicker. The superstitial activates and runs from the browser cache when the Net user clicks on an element on the page.

In an interview with the *Industry Standard* in 1999, Sean Black, vice president and interactive media director at Grey Line Interactive, claims that superstitials have surpassed the ad agency's expectations. According to Black, about 8% of Net users who saw the ad interacted with the content (banner ads have click-through rates of under 1%) (Anderson, 1999). While rich media ads like superstitials cost up to four times as much as a static banner ad campaign, these types of interactive advertisements generate higher rates of interaction and click-through. As with any advertising, rich media advertisements should act as a supplemental rather than a sole revenue stream.

Evaluation Points

Advertising works best when a publisher has the following elements:

- *Diversity in revenue:* Ad revenues are one part of an overall business model. Even Yahoo!, one of the only sites on the Web to turn a profit in advertising, is branching out into different revenue streams.
- *Traffic volume:* Ad revenue lives and dies on traffic volume. Unless your site gets between 8 and 10 million visitors a year, advertising requires a supplemental revenue stream.
- *Demographic statistics*: Advertising rates depend on demographics. TheStreet.com gets much less traffic than a portal site but has strong demographics to drive its high CPM rate of about $55.
- *Conversion rate tracking*: Advertising, like any other revenue stream, needs to be tracked. Metrics include page impressions, click-through rates, and conversion rates. A company's command of its advertisement metrics demonstrates to advertisers an ability to gauge a return on their CPM investment and helps a publisher figure out what advertisements appeal to the target demographic.

Syndication

Syndication is both a distribution mechanism and a revenue stream for content. Content creators in the physical world, such as film studios, journalists, and cartoonists, have streamlined the process of syndication. Television channels syndicate sitcoms, cartoonists syndicate their work to newspapers, and news bureaus syndicate to local papers. Open your city's daily edition and chances are you'll see a byline for Associated Press or other wire service. Syndication allows a content provider to create an item of value and distribute it via multiple channels. It's a natural fit with the Internet. Syndication allows companies to exchange information more efficiently and easily.

National Semiconductor uses syndication to provide technical information updates to a number of OEM partners and distributor Web sites. The National Semiconductor catalog, which includes specification sheets, schematic diagrams, tech notes, and parts lists, runs over 100 MB in size and requires at least daily updates. National Semiconductor uses Vignette's Story Server to distribute incremental changes to OEM and distributor Web sites. Updates are made using automated workflow within National Semiconductor, and the results are pushed to the National Semiconductor corporate intranet and customer Web sites. Syndication allows National Semiconductor to make updates to one set of content and simultaneously distribute those changes to any number of subscriber sites (including its own intranet and corporate Web site).

The Roles of Syndication

Syndicated content flows between content provider (creator of the content) and distributor (host of the content) via a syndicator or a channel set up between two content management servers (see Figure 4.3). Syndication of content enhances any value exchange with the consumer, with the exception of a promotional value exchange.[10] FitForAll (*www.fitforall.com*), a fitness and health portal, uses content syndicated from the Vitamins.com health site (*www.vitamins.com)* to provide in-depth information that the portal doesn't have the resources or expertise to generate. Syndication helps portals such as Yahoo! gain and retain customers by aggregating expert, in-depth information in a single location. Syndication also feeds commerce by providing context for a purchase. For example, an online car dealership can offer a deal on a car accompanied by a syndicated review of the vehicle model from Car and Driver Online. The additional information from a respected third party helps the consumer evaluate and (hopefully for the dealership) consummate the purchase. From the content provider's perspective, syndication allows the publisher to monetize its core competency. TheStreet.com is starting to build up its revenue stream around syndication by providing premium content to both portals and corporate intranets.

TheStreet syndicates features to online brokers such as E*TRADE and TD Waterhouse Investor Services, services such as iSyndicate, corporate intranet

---

[10]Promotional sites do not want to offer content from other sites because it detracts from the focus of the value exchange: promoting the company's core product or service.

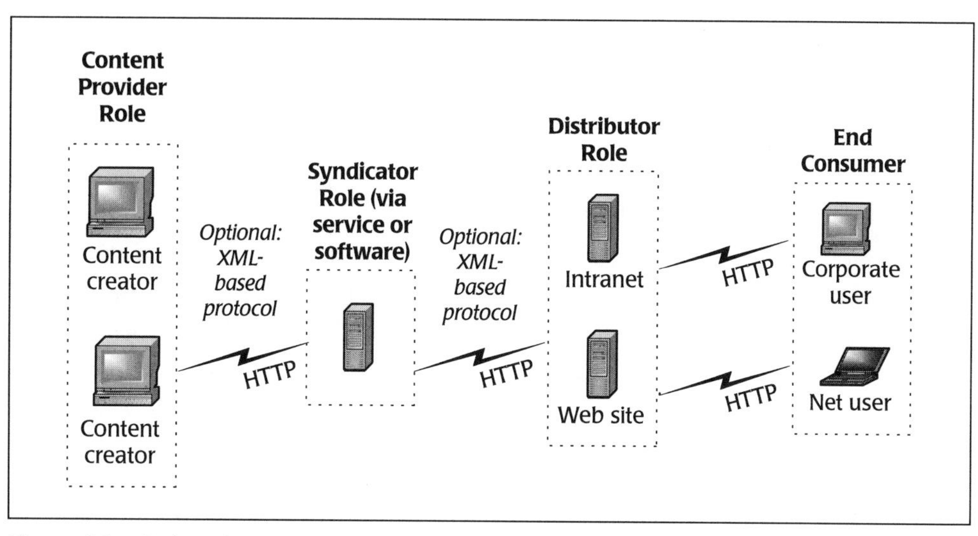

Figure 4.3 Roles of Syndication.

customers such as Fidelity, and Net user portals such as AOL, Yahoo!, and Infoseek.

Syndication Flexibility in Delivery

"Make content relevant and deliver it in context" is the maxim that Joel Maske, the innovator of online syndication services on the Web, preaches to both syndicators and content providers. Maske's company, iSyndicate, has been setting up and maintaining syndication relationships since 1996. Maske says that back in the early days, his company's chief challenge was educating the market about the possibilities of online syndication. Many content providers were receptive to the idea of multiple channels of distribution. The benefits for them are concrete. Content providers earn a fixed licensing fee for syndication as well as any additional lift in revenue that can be attributed to the syndication. iSyndicate acts on behalf of the content partner to distribute the content. It dynamically updates syndicated content on remote publisher host sites and splits the revenues from licensing with the content provider.

iSyndicate's service handles syndication of either the entire content or "teasers" (generally headlines or abstracts) that lead back to the content provider site. The company's Express product syndicates free headline links as teasers to attract Net users to a publisher's site. The publisher of the free headline links pays iSyndicate 2¢ per click-through for the increase in site traffic (J. Maske, phone interview, 20 Aug. 1999).

Syndication as a Community Enhancement

The two largest national online community sites (Yahoo! GeoCities and TheGlobe.com) use syndicated content as a hook to gain more members. These community portals partnered with iSyndicate to allow their members access to free headline content from a variety of content publishers (including wire services) on their home pages. As a result of the partnership, a GeoCities member can integrate professional, syndicated content into his or her Web page. Headlines range from top news to comic strips to weather forecasts.

Maske is confident that syndication is the new model for driving traffic on the Web. He sees more portals, such as Yahoo! and Excite, adopting syndication as part of their value exchange for users. "The benefit of syndication," explains Maske, "is the ability to be solely focused on your own domain." By syndicating high-quality content from a variety of sources, portals and other content aggregators are "buying into what you can't do yourself." Maske sees the greatest challenge in syndication in figuring out which content to syndicate to which partner for the greatest value.

Syndication between Businesses

David Mathison, chairman, CEO, and cofounder of the Kinecta Corporation (formerly ShiftKey, Inc.), has been building solutions for content exchange between online business partners for quite a while now. Before cofounding his San Francisco–based start-up, Mathison served as vice president of Global Syndication for Reuters News. From 1994 to 1999, he created and implemented profitable Internet-based products for the giant news wire service.

Mathison's start-up company, Kinecta, offers the Kinecta Interact solution. Content providers use the Kinecta Interact platform to create their own syndication networks, which automatically distribute any digital asset to hundreds or thousands of affiliates and manage it. Content can include catalog data, text, photos, graphics, or streaming audio and video formats. The content provider, not an intermediary, maintains direct control over its subscriber/customer relationships. The Kinecta model allows content providers to handle their brand management, subscriber license management, the placement of contextual advertisements into content, and digital rights management. As a result of "owning" the relationship with their subscribers, content providers can easily deploy existing and new digital products.

Kinecta's customers, such as Reuters, Motorola, iVillage, Fidelity, and TheStreet.com, are among the Web's largest content providers and aggregators. The Kinecta solutions are used to manage distributed information on commercial Internet sites as well as corporate intranets and extranets. TheStreet.com, for example, uses Kinecta to syndicate content to the Fidelity intranet. With high-volume customers such as Reuters, Kinecta has dedicated considerable resources to developing the scalability of its systems. Mathison explains that in 1999 Reuters alone provided 150,000 news stories per week using the Kinecta Interact platform, including most of its wire services in English, Spanish, and fifteen other foreign languages.

Both Mathison and cofounder and chief strategist Adam Souzis (formerly engineering lead at NetObjects) see syndication and distributed information management as an ideal fit for Web business relationships that leverage multiple points of presence at portals and intranets. "The real action," says Mathison, "is going to be in the business-to-business arena. Almost all online businesses have a need for an easy, yet robust and scalable, way to exchange digital assets with partners" (D. Mathison, phone interview, 10 Nov. 1999). Souzis adds that corporate portals and intranets are getting increasingly

Protocol to Watch: ICE

The Information Content Exchange (ICE) standard, which is up for review by the World Wide Web Consortium, defines a new channel for information exchange. Based on Extensible Markup Language (XML), ICE lets content providers and publishers exchange content directly between servers without requiring an intermediary third party. ICE is supported by companies such as Adobe, Kinecta, Microsoft, Sun Microsystems, and Vignette as the preferred protocol for information exchange. Version 1.0 of the ICE specification was realized in October 1999. It defines the XML-based syntax for syndicator-to-distributor transactions.

sophisticated and need to create and support correspondingly more sophisticated content relationships (phone interview, 10 Nov. 1999).

Kinecta's solutions are compatible with the ICE protocol governing the syndication and management of digital assets and are fully interoperable with other ICE-compliant servers, such as systems from Vignette. In fact, Souzis was intimately involved in defining the ICE standard for Web-based content syndication as a member of the ICE Authoring Group. The emphasis on scalability, infrastructure, and open protocols makes a syndication system such as Kinecta Interact appealing to content providers interested in establishing a robust syndication infrastructure within their production environment.

Evaluation Points

Syndication works best when a publisher has the following:

- *A clearly defined context for information*: With syndication, it's generally up to the distributor Web site where and in what context to place the feeds. Make sure that you're comfortable with how and where your content may end up on the distributor Web before you go into syndication. Even though the content is not on your Web site, it's still your brand.
- *Limited site porousness*: If you license your content, the distributor hosts the content on the remote site. That means your visitors might be going to the distributor's site to get their information rather than yours. Plan syndication products carefully to ensure that you aren't licensing away your most valuable content (unless syndication is your sole revenue stream).
- *A way to manage multiple relationships*: Whether you outsource your syndication distribution relationships through a service or conduct them in-house using syndication solutions, you must have a scalable way to manage the relationships. Since the relationships involve a licensing fee and near-real-time delivery of content, it's important to maintain an efficient process for distribution and revenue collection.
- *An open mind about syndication*: Any digital information can be syndicated, including articles, photos, music, streaming media, and text.

Pay per Access

Pay per access is the online version of pay-per-view movies on cable. The Net user provides payment in exchange for viewing content, downloading content, or using content privileges. Pay per access for premium content provides the greatest flexibility in price for Net users, because it allows the publisher to set up a single item or bundled products. A "product" consists of one or more files (which can be a Web page, an MP3 file, or a streaming media feed).

Complexity in Product Composition

A product can be simple or complex. A simple pay-per-access product consists of one file. Digital image provider Corbis sells screensavers of high-quality images in single executable files, which, when run on the buyer's machine, install the screensaver. A complex product consists of more than one page or file packaged as a single file. For example, Corbis can bundle three screensavers (three executable files) as a single product and charge a higher rate.

Pay per access provides the greatest flexibility in merchandizing for publishers and purchasing for consumers. For publishers, pay per access allows for a wide range of pricing strategies, ranging from micro-payments (such as the *Wall Street Journal*'s 75¢ daily edition) to higher dollar values. Net users benefit by having the flexibility of buying access to files on a per-item basis or bundled as a single product. The bundling opportunities also allow the Web publisher to use promotions and other marketing techniques, such as cross-selling and up-selling, to promote other related products.

Buying à la Carte

Pay-per-item access to content allows consumers to fulfill their objectives without spending time or money on unnecessary content. For example, *The New York Times on the Web* uses pay-per-access pricing for its archives and crossword puzzles. The Web site lets you run a search through its extensive archives but asks for payment when you click the headline link to get to the text. Researchers and other users of the *Times* archives have the flexibility of buying articles "à la carte" without paying a subscription fee for the entire site. In addition, since many subscription services automatically bill a credit every month until explicitly canceled, pay-per-item pricing lets users engage in a finite commerce relationship around content. The user is simply charged for the single access (which can have a duration of an hour or several weeks).

Handling Digital Content Sales

Since e-commerce servers and content management systems do not support pay-per-access payment structures, services such as Qpass (*www.qpass.com*) provide publishers with outsourced tools and transaction handling for the per-item charges. Qpass, which handles the pay-per-access pricing of the *New York Times* archives, launched the content commerce service in 1999 with pay per access as one of several content-selling models. (Other payment models include subscription and point-of-sale syndication.) Qpass establishes the billing relationship with the customer, aggregates single-item content transactions, and then hits the customer card for payment once a month. Publishers share a percentage of the content sale with Qpass in exchange for the service's transaction aggregation, customer account management, and customer services. So, for example, if a customer has a problem downloading a Corbis screensaver, Qpass would handle the customer support call, including refunds and credits to the Net user's digital content account. Publishers use the service to package content and associate properties, such as price and promotions, with the product. Qpass also offers sales and revenue reporting to help publishers track the success of their products.

Bill Bryant, cofounder and chairman of Qpass, points to the success of pay per view in the cable world when explaining the possibilities for this type of content commerce. He sees the merchandizing of content eventually evolving into "business models that parallel cable: a bundle of offerings for a set price, with additional premium services offered on a pay-per-view basis." Bryant, who spends a considerable amount of time educating content publishers about the revenue possibilities of content commerce, sees pay per view being adopted when publishers gain a greater understanding of how to "merchandise, market, and promote their [content] offerings" (B. Bryant, interview with author, 29 June 1999). Services, such as Qpass, give publishers the ability to package and sell content as a commodity and provide much needed flexibility to pricing models on content and entertainment sites.

Evaluation Points

Pay per access works best when a publisher has the following:

- *Granularity in content*: Individual articles, broadcast feeds, and screensaver executables all constitute content that can be easily broken into individual products.

- *Merchandizing flexibility*: Pay per access is especially amenable to promotions like two-for-one deals or buy-three-get-one-free. This pricing model can be used to sell a single item or to bundle multiple items to raise the value of the product in the Net user's perspective.

Subscription

Subscription-based access to premium content works to monetize content value exchange when the information accessed is highly differentiated from the free content. Every site should offer at least some free content. Without a certain amount of quality free content, the publisher loses the ability to build out revenue streams in addition to subscriptions. Forrester Research's 1999 study on content commerce ("The Content Commerce Collision") notes that subscription-based models will see the slowest growth in the next five years, with the bulk of fees being generated by business archives and adult content.[11] Pornography sites generate the most subscription revenue today. In 1998, adult content drew 69% of the $1.4 billion market for paid content on the Internet. Analysts predict that by 2003, content commerce will grow to $5.2 billion—with the adult content share dropping more than 10%, to only 58% of the total market. Targeted areas for growth include computer games, premium sports content, music, and other digitized information (Charron, 1999).

Degree of Relevance in Subscriptions

The subscription-selling model is successful in two contexts:

1. *Diversification:* As one of several revenue streams for a high-quality content provider.
2. *Cross-channel sales:* As an online demonstration of the print product's value or promotional pricing for print/online subscribers.

TheStreet.com demonstrates the benefits of *diversification* in a business model. TheStreet sees about one-third of its revenues from subscription content, with the remainder coming from a mix of advertising, sponsorships, and syndication. The publisher's base has risen 40% each year since the site

[11]Today, a full 70% of current subscription fees fall into the adult-entertainment category. Information from "Paid Content on the Net," 1999 (*www.emarketer.com*).

launched, with an impressive renewal rate of more than 90% for the annual subscription (sold at rate of $100 per person). TheStreet would not have been able to build its subscription base without providing unique and high-quality content as part of the value exchange.

The Wall Street Journal Interactive Edition (WSJIA) provides the most successful example of cross-channel sales. The *WSJIE* subscription site generates revenue because the publisher provides reduced pricing for existing print subscribers and bundles considerable high-quality content for $59 a year, including access to the financial magazine *Barron's Online,* online portfolio tools, and access to the extensive *WSJIA* archives. In addition, the Web site acts as a sales tool for the print *WSJIA*, encouraging Net visitors to pay the higher cost for the printed publication.

To monetize value exchange with subscription, the publisher has to be very certain of the value proposition to the Net user. Both TheStreet.com and *The Wall Street Journal Interactive Edition* work hard to maintain a high quality of content and unique value for their subscription users.

Test-Driving Subscription Value

If the publisher produces content that has less relevance to the consumer (i.e., doesn't drive critical decisions or assist highly prized hobbies), the publisher can run a test to gauge the Net user's reaction to a subscription model before implementing it. Subscription content can be packaged up as low-cost pay-per-access content for a short time to gauge the demand. A high demand indicates that subscriptions, sensibly priced, will be successful. A low demand rate means that Net users don't value the content enough to pay for it. In that case, a publisher is often better off bundling multiple pieces of the content together in a pay-per-access offering. Bundling lets the publisher aggregate value until the product presents an attractive enough value proposition for consumers to buy.

Evaluation Points

Subscriptions work best when a publisher has the following:

- *Multiple files in a single "product"*: The publisher simply updates files for a subscription product, while product attributes (such as price) remain the same.

- *Flexible time increments*: Subscriptions should be offered in weekly, monthly, quarterly, annual, or other time intervals. Ideally, a subscription service also offers a trial period.

Secure Distribution

Secure distribution of content opens up the ability for high-value information to be displayed and sold over e-mail as well as the Web. It's no secret that secure distribution technologies are complicated pieces of software. InterTrust and Xerox both spent years developing the infrastructure to support systems that check for the user's rights before allowing access to content. Publishers have the option of going directly to these system providers and licensing their software (an expensive proposition), buying a publishing system that offers digital rights management (such as Xerox's Content Guard), or outsourcing the distribution to a service bureau.

PublishOne, a young company based in Santa Clara, California, provides one such service for high-value content. Founded in early 1999 by CEO (and former founder of the Internet Shopping Network) Kirk Loevner, PublishOne offers a shared revenue business model for its customers. The PublishOne system is built on top of InterTrust's MetaTrust technology, which provides the architecture for the access management.

Using a service bureau like PublishOne simplifies the packaging process for digital content. Publishers navigate to the service's Web site, indicate which files belong in a given product, and set product attributes such as price and promotional information. The PublishOne system creates a DigiBox (InterTrust's secure format) that contains the documents that belong in the product. When publishers offer the content for purchase on their Web sites, they provide a link to download the DigiBox file format.

A consumer downloads the file in the secured format and is prompted for payment when opening the file for the first time. (The consumer must have custom software on his or her machine before opening the DigiBox file.) Every time the file format is opened, the authorization software automatically checks to see if the consumer has already paid before permitting access. Secure distribution opens up the opportunity for Web publishers to sell their content to a large number of people without diluting the value of the product through unauthorized distribution (see Figure 4.4). Publishers collect payment every

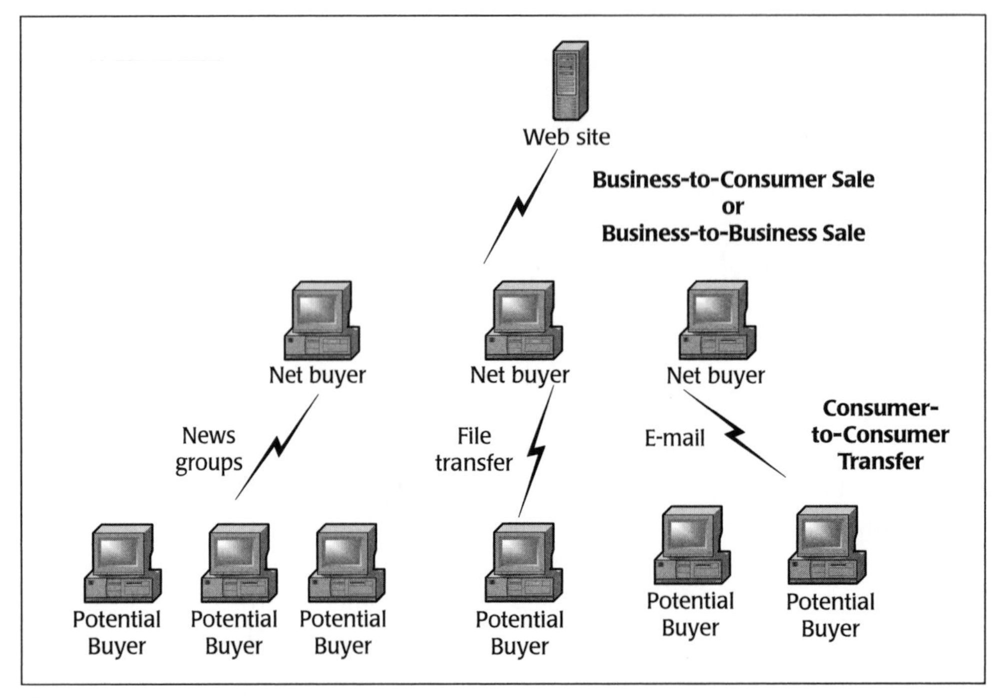

Figure 4.4 Secure Distribution.

time the content is accessed, whether the Net consumer obtained the file from the publisher's Web site or as an e-mail attachment from a friend.

Secure Distribution as a Business Model

When describing secure distribution as a revenue stream, Loevner's careful to point out that digital rights management is not about protecting what you own. "Think of it as the ability to expose the content that you weren't able to put on the Internet or in e-mail before," Loevner says. "It's far more about *enabling* content distribution than restricting it" (K. Loevner, phone interview, 8 June 1999). He sees considerable interest in the secure distribution model from vertical industries, such as legal information, and research firms who may or may not have a strong Internet presence today. The model of secure distribution addresses the question, "How much would it cost you if this information was not secured?" as much as, "Have you paid for this information yet?" Both questions are equally applicable for any high-value information sold through e-mail or the Web.

Loevner explains that secure distribution technology allows publishers to apply very specific rules to the usage of the file. For example, with his company's secure distribution, publishers can indicate that a file can be viewed but does not support cut-and-paste or cannot be printed. Not only can you define this level of granularity, but Loevner says that you also can track the exact time and type of usage.

This type of control over the usage of content makes a secure distribution mechanism attractive for users of Internet services such as e-mail, as opposed to just Web shoppers. Right now, many corporations and almost all lawyers do not risk sending sensitive information in e-mail because of the chance that the e-mail could be forwarded and the information spread to unauthorized parties. Let's consider a case in which the CEOs of two companies have begun to discuss a multimillion-dollar merger. The top executives of the oil companies need to communicate sensitive details without having any leaks to the press (and to shareholders) until farther along in the process. Any leak of the impending merger would have a disastrous impact on negotiations and shareholder response to the event. If contracts and other legal documents are shipped back and forth between the companies as simple Word files, any unscrupulous employee could distribute that sensitive information. A secure distribution method lets the executives create a package of information that can be accessed only by the people that the executives specify. That way, if an e-mail does get forwarded with a secured file, unauthorized people are not able to open it.

Evaluation Points

Secure distribution works best when the following is true:

- *Content has high value*: Content is unique or not easily available through the Internet or other channels.
- *Custom software is easily distributed*: Secure distribution technologies depend on custom software on the Net consumer's computer to enforce business rules (like payment). In order to allow any access (authorized or not) to the content, you must provide the custom software to users to install on their computers.

- *Reporting is highly detailed*: Since the custom software manages access, secure distribution systems have the opportunity to collect fine-grained detail about usage of the content, such as the number of times content is printed.

Summary

Net consumers' demand for different forms of value exchange has sparked an evolution in revenue streams. "Traditional" e-commerce models consisting of fixed-price products have given way to more flexible pricing and new forms of consumer-driven merchandizing and inventory. The success of auctions and consumer merchandizing demonstrates how Net user goals of access to community and low cost helped develop new variations of the traditional commerce model.

The evolution has been the integration of content and commerce. New technologies supporting flexibility packaging and payment options give publishers the ability to treat high-quality content as a product. And content providers, such as TheStreet.com, *The Wall Street Journal Interactive Edition,* and *The New York Times on the Web,* are leading the way in experimenting with the new models for commerce around content exchange.

The merchandizing of content creates a new marketplace that supports multiple revenue streams around digital goods. Content publishers no longer have to depend on advertising, which for all but the most highly trafficked sites does not provide substantial revenues. Instead, content partners can earn dollars from multiple sources (syndication license fees, pay-per-access content, subscriptions) in addition to advertising to form a *hybrid* revenue model. These revenue models provide the first real opportunity for content publishers to earn money from their core competencies of creating and aggregating quality content.

PART TWO

Strategies

Part One focused on identifying the characteristics and business models for killer content and services. Part Two focuses on *strategies* for successful value exchange—specifically, the features that enable consumers to interact with and react to killer content. It does not attempt to provide step-by-step instructions on building a retail Web site, a content portal, or the transaction processing systems that support a commercial Web site. This part looks at three strategic areas that support value exchange around content:

1. *User experience* while browsing and interacting with content on the site (chapter 5)
2. *Site structure* required to support the user experience (chapter 6)
3. *Internal processes*—the workflows and relationships in the production environment that support the user experience and site structure (chapter 7)

These areas differentiate Web sites according to the needs of target consumers. As a result, you may find that some of the strategies offered here don't apply to your Web site. For example, a vertical portal that syndicates premium content from other sources does not need digital rights management to protect against unauthorized redistribution of the content; after all, the vertical portal does not own the content. The provider who created the content, on the other hand, will be interested in understanding about rights management, especially if the content has a perceived high value or can be easily redistributed in file format.

Other strategies apply more universally to Web sites. For example, clearly stated and easy-to-use options for customer communication benefit all Web sites, whether the site creates or aggregates content, sells retail goods, promotes a company, or offers entertainment. Amazon.com depends on the customer loyalty fostered in part by effective customer communication for repeat purchases. Entertainment content site MP3.com uses customer interaction to gauge the popularity of bands and consumer preferences for music. When reviewing these strategies, keep in mind what you offer to consumers as part of your site's value exchange.

The chapters identify strategies that enhance value exchange, so that you can take the next steps of evaluating the cost of implementing the solution versus its benefits. Since the steps to that cost analysis differ radically by company, we will look at the benefits for each strategy. Ultimately, you are the one who can identify and prioritize the right strategies to enhance your site's value exchange around content.

Picking the Right Strategy for Your Site

The strategies in Part Two range from techniques for interacting successfully with consumers to the importance of intelligent cataloging of content for personalization. Implementing every single suggestion would be expensive, both in time and resources, and probably end up detracting from your goals in value exchange by overloading the user experience and your ability to support it. The key to picking the right strategy for your site lies in the Net user's goals and expectations in your specific value exchange. Remember that a successful value exchange occurs when the Net user is able to accomplish his or her objective at your Web site and has a smooth user experience in the process. When evaluating the right feature to implement on a Web site, focus on the strategies that support the Net user's value factors, which were discussed in Part One: *utility, relevance, timeliness, innovation,* and *credibility.*

To help with that evaluation, each strategy mentioned in the next three chapters is summarized in a sidebar, which includes the key business driver for the feature, the effect on the value exchange, the impact on the structure of your Web site, and ways to maximize the value exchange. The summary section also contains a list of Web sites or third-party vendors that do that particular aspect of the relationship or infrastructure noticeably well. Whatever feature(s) you choose to implement should fit within the site framework.

The user experience and site infrastructure directly support the consumer's interaction (communication) with content and the publisher. The internal processes within your company's production environment enable a superior user experience. We will begin by discussing strategies for supporting and enhancing the user experience.

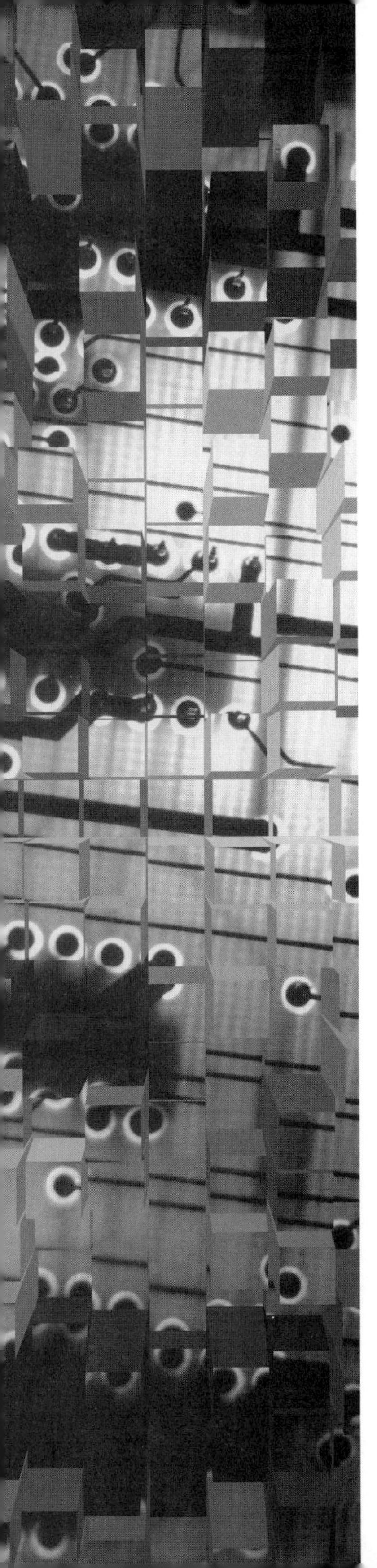

CHAPTER FIVE

Supporting User Experience

Successful value exchange cannot occur if a Web site ineffectively frames the exchange between Net users and its publisher. A Web site's framework—the features and processes that support content and commerce exchange—provides the context and tools for a Net user to engage in a value exchange.

All commercial Web sites pay at least some attention to user experience and site layout. Even the *International Herald Tribune*'s Web site provides the most basic elements of a framework: a home page, secondary site pages, and a search facility. However, consumers can't easily interact with the content by locating stories or viewing images that complement a story line. As the *IHT* site demonstrates, the lack of a proper site framework inherently diminishes the Net user experience.

This chapter focuses on how to build out the relationship between consumer and Web site. It focuses on strategies and does not go into the details of how to build a Web site, offer layout suggestions, or include HTML coding tips. If you're interested in creative suggestions for improving site layout, David Siegel's book *Creating Killer Web Sites* provides solid advice (Siegel, 1997). For coding tips for lightweight but elegant HTML code, Ardith Ibanez and Natalie Zee's book *HTML Artistry* explains the concepts well and offers interesting examples (Ibanez and Zee, 1998).

Here we focus on features that improve the *context* in which content is presented to Net users. Table 5.1 summarizes strategies to enhance a site's user experience discussed in this chapter.

Table 5.1 User Experience Enhancement Summary

Strategy	Description	Benefits
Provide Membership	Offer registration at the site to access premium content and services	Community building, targeted marketing, subscription opportunity (if appropriate)
Personalize the User Experience	Present content in which the site visitor has indicated interest through previous browsing or member profiles	Community building, targeted commerce offers, customer and site loyalty
Support Users	Provide responsive and convenient customer service	Community building, customer and site loyalty, repeat purchases
Communicate via Community	Allow visitors to communicate with each other and the publisher through the site	Community building, customer and site loyalty
Reward Net Users	Provide visitors with rewards for visiting and using the Web site	Customer and site loyalty, promotional product upsell and cross-sell opportunity
Market Effectively	Promote the site's content and products without alienating current and potential customers	Customer and site loyalty, promotional product upsell and cross-sell opportunity
Set Up Smart Affiliate Relationships	Establish affiliate relationships with both private (consumer) and commercial Web publishers	Customer and site loyalty, new revenue stream

Provide Membership

"Membership," as the American Express slogan goes, "has its privileges." Providing exclusive benefits for Net users who trust a content publisher with their personal information is an essential part of content and entertainment value exchange. From a business perspective, a content publisher aims to attract and maintain the ideal demographic visitor and drive purchases. The most effective way to build the right user base is to offer membership that has tangible privileges—even if the membership service is free. Think of membership as a gateway to added services. For a travel site, such as Travelocity.com, membership means the ability to store travel preferences and receive promotional specials for specified vacation spots. For a sports site, such as ESPN.go.com, membership grants users access to exclusive commentary, editorials, and other content from respected sports columnists. To be effective, member services and content must provide a clear benefit over the material available to casual visitors on your Web site. Providing the same level of content and no added services to members eliminates incentive to join. A membership not valued by the user is a membership not used after the first investigative visit and an opportunity lost to the content provider.

Acquiring Information for Lost Passwords

Membership on Web sites should require at minimum a name and e-mail address. An address provides an avenue for communication with the visitor. If there is a clear delineation between "free" content (available to any visitor to the site) and "premium" content (requiring membership to access) on a Web site, the publisher also needs to ask for a user-defined password.

As a rule, it's safe to assume that every password will be forgotten at least once by the visitor. To save support costs, ask for additional user identification so that you can do an automated lookup of the password. That information can be customer-specific (such as a birthdate), used in correlation with a member identification number, or a password hint that allows the user to remember his or her own password.

Some sites (such as Yahoo!) offer both types of password lookup and give the customer the option to choose the preferred method. When asking for a password hint, many content publishers include a common field such as mother's maiden name. However, it's more secure to ask the visitor for the hint that he or she prefers.[1] Publishers have the option of displaying the password[2] on the computer screen after lookup or e-mailing the information to the consumer.

Web Site Cookies

Many sites use cookies to store data such as member name and password. A cookie is a small text file stored on the user's computer that contains information such as a user's name. When a user arrives at a Web site, the site recognizes the cookie, extracts the user name and password, or performs some other action based on information in the cookie. Cookies contribute to a more streamlined user experience by quietly passing information on the visitor's behalf to a Web site. However, users can configure Web browser settings to not accept cookies from Web sites or to prompt when downloading the cookie. If you plan to use cookies to streamline authentication, make sure that you provide a good user experience for visitors who do not accept cookies or who use older browsers that do not support cookies.

Acquiring Information for Marketing Products

Most membership forms request other personal information, which may include personal preferences and purchase information. On an Internet that hasn't quite figured out what to do about privacy, Net users are selective about parting with personal data. In return for a Net user's personal information, a content publisher should provide benefits to the Net user and explain what the benefits are. Travel site Travelocity tells Net users that entering personal information, like address information and preferences for air travel seating, helps the Net user book tickets more easily at the Travelocity Web site. Content commerce provider Qpass tells potential members that providing address and payment information to the registration service results in the ability to use an automatic form-filler utility. Getting members to register means making it worth the Net user's while

---

[1]It's relatively easy to find out a Net user's mother's maiden name, user name, and e-mail address. If the hacker knows the user's name and e-mail address, he or she has the ability to "request" a password reminder by offering your name and mother's maiden name as proof of identification. To avoid this risk, let the visitor define the hint for a password.

[2]This, of course, cannot be done in cases in which the member forgets his or her member name, e-mail address, or the answer to the password hint.

to table doubts about privacy and take the time to fill out the forms. From the publisher's perspective, membership paves the way to future success because it opens the door to data mining.

Obtaining information from a member base translates into two benefits. First, the publisher can better target content using the demographics of the member base. If half the member base is from San Jose, California, the publisher can target its content creation budget at building out the tech-friendly parts of the Web site. Or if 75% of all members click on or buy syndicated content, the publisher can increase the amount of syndicated information to the Web site.

Second, data mining sets up better, more profitable relationships with affiliate and advertising partners. A potential partner will be impressed if a content publisher can (reliably) claim that 50% of site traffic stems from site members for whom credit card information and some demographic data are on file. While many privacy policies forbid sharing specific member payment and purchase information, a content publisher can analyze individual activity to come up with aggregate information (such as "50% of my Web site's members purchase here at least once a month"). Statistical information about members and traffic drives up value with partners and lets a content publisher track the bottom line against business objectives.

Strategy Summary: Membership

Business Drivers: Community building, potential to market products or content to the consumer based on membership information, opportunity to introduce a subscription payment model (if appropriate)

Value Exchanges: Content, entertainment, commerce

Site Impact: Requires restructuring of Web site into free and access-protected areas, and setting up authentication (login) for when a Net user tries to access protected content

Content Business Models: Pay per access, subscriptions, syndication

Best Examples: Travelocity (*www.travelocity.com*), *The New York Times on the Web* (*www.nyt.com*), TheStreet.com (*www.thestreet.com*)

Personalize the User Experience

Personalization means the publisher's ability to gear content on a Web site to individual tastes and preferences. A personalized experience for the Net user results in the custom delivery of content based on user-defined preferences. From a business perspective, personalization develops Web communities. Net users like personalized Web sites because they have control over the content. Yahoo! provides a personalization service called My Yahoo! (see Figures 5.1 and 5.2). To get this service, a Net user registers with the site, providing name, e-mail, and password.

My Yahoo! offers the Net user a personalized view of all the information on the Yahoo! site, including access to weather, message boards, and chat rooms, and even includes personal information management features, such as an address book or calendar. In addition to content, the personalization service provides the ability to modify the color schemes and even change the layout of

Figure 5.1 Home Page for Yahoo! (without Personalization).
Source: Copyright © 1999 by Yahoo!, Inc. YAHOO!® and the YAHOO!® logo are trademarks of Yahoo!, Inc. Reproduced with permission.

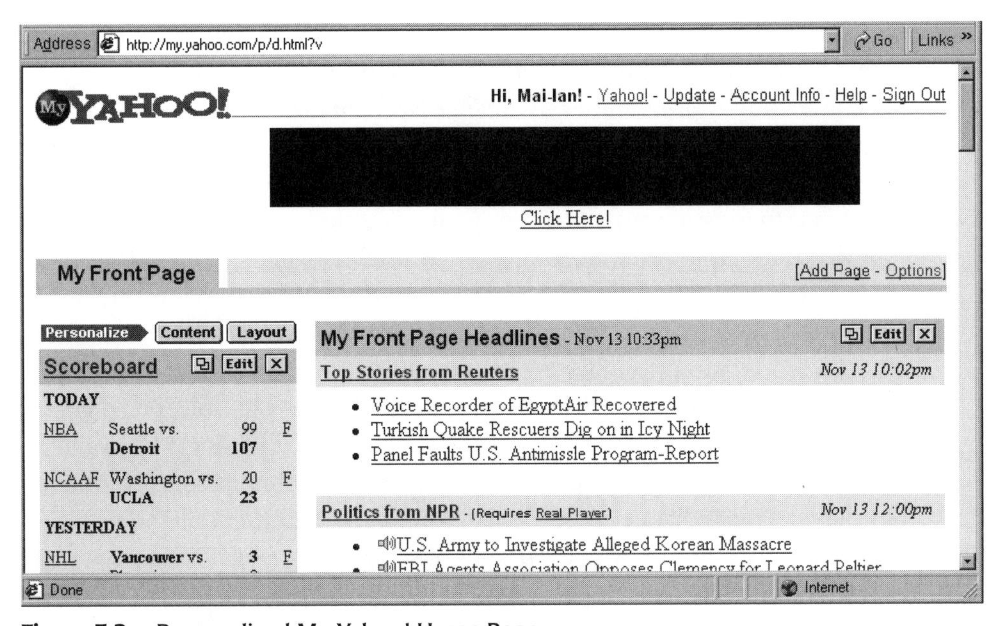

Figure 5.2 Personalized My Yahoo! Home Page.
Source: Copyright © 1999 by Yahoo!, Inc. YAHOO!® and the YAHOO!® logo are trademarks of Yahoo!, Inc. Reproduced with permission.

the site. Layout changes allow you to rearrange how information is displayed on the page. For example, you can move the stock ticker to the top of the page and weather updates to the bottom of the page, if the first items you want to see are your favorite stocks.

The appeal of My Yahoo! lies in its convenience. Net users select content on the portal that specifically interests them. Rather than spending time browsing different channels on Yahoo.com, Net users go directly to their personalized pages and get the information they need in a single trip. Convenience and control over content provide compelling reasons for a Net user to set up a single personalized vantage point and stick with it.

For the publisher, personalizing the Net user experience provides two benefits: the ability to attract and keep new members and the ability to monitor user preferences. Personalization is a choice service gained through membership. The more content and control granted to Net users, the more popular the personalization service. In addition, think of the indirect feedback that members are providing through their personalization choices.

A content publisher can track popular categories of content by monitoring what members choose for personalization options. What percentage of Net users builds predominantly financial personalization pages? What percentage prefers leisure topics, such as sports or gardening? This data gives you the tools to make business decisions about where to spend your content creation, partnering, advertising, and syndication dollars. A 1999 Jupiter study showed that out of sixty merchants, 40% used personalization at their Web sites (see Figure 5.3). Of the 60% not currently using personalization, a full 93% were planning on integrating personalization in the next twelve months.

Personalization provides benefits to content, commerce, and entertainment value exchanges. As described in the My Yahoo! example, personalization on content sites gives Net users the ability to set up a single, customized vantage point for information. Similarly, personalization on an e-commerce site lets Net users dictate exactly what products interest them on the Web site. For catalog-driven sites with thousands of products, personalization gives Net users the ability to organize their own "store." Take the personalization offered by Bluefly (*www.bluefly.com*), an online designer clothing outlet. Bluefly provides a customized MyCatalog (MyBluefly doesn't quite connote the desired image). The MyCatalog personalization service lets Net users set up their sizes, favorite designers, and preferred types of clothing.

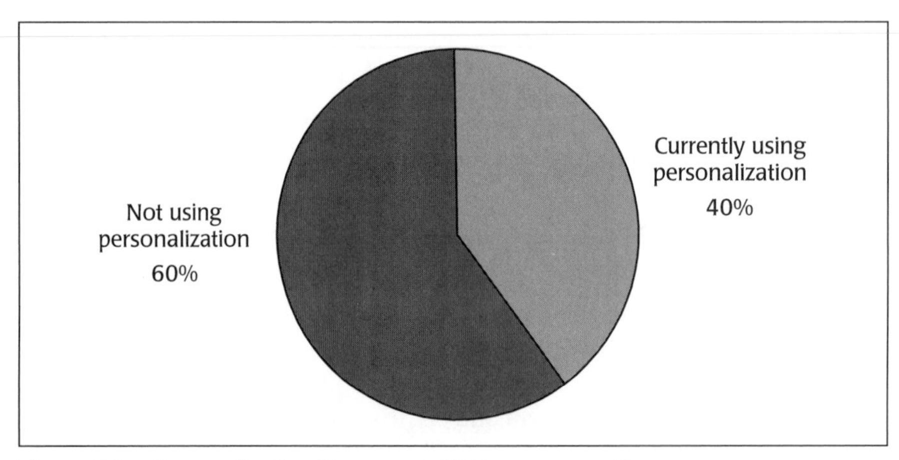

Figure 5.3 Personalization Use across 60 E-commerce Sites.
Source: From "Suggestive Selling Online," Jupiter Communications Online Intelligence, New York: jupdata 2Q1999. Copyright © 1999 Jupiter Communications. Used with permission.

To use the service, Net users set up their name and password, and optionally, an e-mail address. When Net users view their personalized catalogs, only products that fall within the selected categories are displayed. Bluefly offers thousands of products for sale. Browsing through 150 thumbnail images of jackets takes time and patience on a modem connection. The number of images on the clothing Web site makes Bluefly one of the slower catalog e-commerce sites. With MyCatalog, you have access to a personalized view of Bluefly's inventory and a more streamlined user experience while at the Web site.

Bluefly takes the personalization service directly to e-mail as well. When Bluefly gets a product that matches the preferences in MyCatalog, it sends an e-mail to the Net user with that information and a link to the new product. The level of service almost equals that of a personal shopper, who calls you at home when something you might like arrives at the store. "Push" personalization, in which the publisher sends information about personalized content or products to a Net user, builds customer loyalty and drives transactions. It also requires that a Web site integrate e-mail service with personal profiles and catalog data, which can be a time-consuming operation. "Pull" personalization, in which a Net user finds information by browsing a personalized page, takes considerably less work for a publisher to manage and still provides an accessible and popular service for Net users.

Strategy Summary: Personalization

Business Drivers: Community building, targeted commerce offers, customer and site loyalty

Value Exchanges: Content, commerce, entertainment

Site Impact: Requires marking all potential content served to a Net user according to the consumer profile

Content Business Models: Targeted advertising, subscription, pay per access, syndication

Best Examples: Yahoo! (*www.yahoo.com*), Bluefly (*www.bluefly.com*), Infospace (*www.infospace.com*)

Support Users

The Internet is a democratic forum. Interaction plays a significant part in the value exchange. Net users enjoy the freedom to comment on a product, discuss an article, or complain about content. The lowest common denominator for Net user interaction is customer contact. As a rule, every single Web site—regardless of value exchange—should offer the ability to interact with the publisher through e-mail, chat rooms, or bulletin boards. If a content publisher's Web site includes any type of interchange that could present a Net user with difficulty in fulfilling goals and expectations, the site *must* offer prompt help. On a content-pure site, contact information can be as simple as an e-mail address or phone number. Retail commerce sites require a much more involved customer service framework, capable of tracking order status and providing refunds and returns.

Customer service represents a basic courtesy to Net users that essentially tells visitors that the publisher values their time. Similarly, the speed of reply matters. A fast response to a question indicates the desire to help Net users if they encounter problems with membership or purchases. Remarkably, customer service, even at the most basic level, is not yet a standard on the Net. A Jupiter study looked at the response rates for customer service at 125 Net user Web sites, including shopping, travel, financial service, Net user brand, and content sites (see Figure 5.4). The methodology was straightforward. The analyst submitted requests for help via e-mail and logged the response rate for each Web site. The study showed that almost 35% of the sites responded within one day, and 31% took more than a day. A full 24% of the sites failed to respond at all (Foster, 1999).

Along with maintaining a general impression of responsiveness, proper customer service helps Net users make the leap from a browser to a purchaser. According to an NFO Interactive Survey, 35% of online shoppers surveyed would buy more products via the Internet if they could speak to an actual person at the time of purchase. In addition, 13.7% of casual visitors surveyed would make their first purchase if they could communicate with a customer service representative. Such statistics provide evidence of the need to offer clear channels of support for Net users who encounter usability or quality issues on a site.

The best and worst customer support systems on the Web today service retail Web sites that do commerce value exchange. Since the quality of cus-

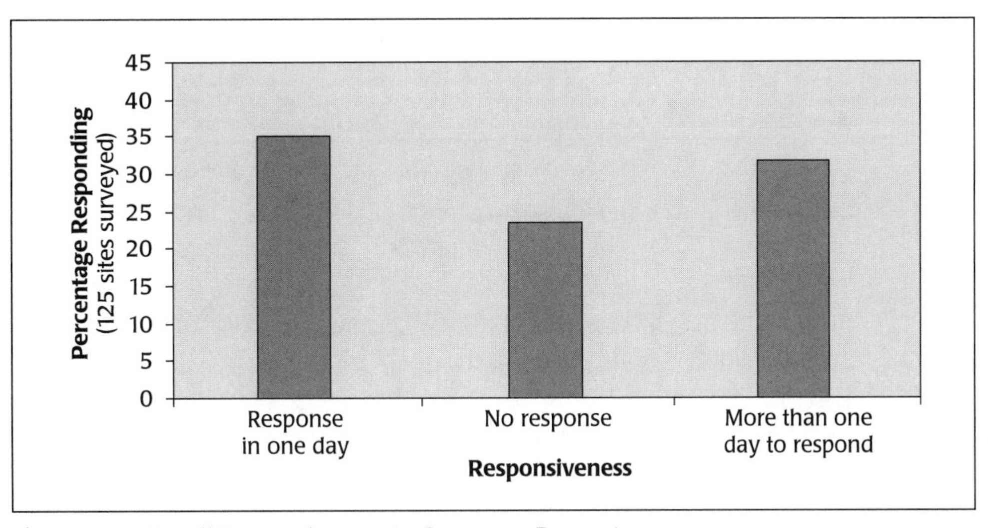

Figure 5.4 E-mail Responsiveness to Consumer Requests.
Source: From "Email Response Rates," Jupiter Communications Online Intelligence, New York: jupdata 2Q1999. Copyright © 1999 Jupiter Communications. Used with permission.

tomer support impacts the likelihood of repeat purchases, smart commerce sites spend time and money making sure that they can respond to order tracking and billing disputes. Amazon.com does customer service especially well. It frequently provides free upgrades for delivery service and offers customers the ability to track orders. Customers can then send e-mail or call a service representative to discuss an order. Self-service tools for tracking purchases on the Web site combined with a lenient refund policy have contributed to Amazon.com's success as a Web retailer.

In contrast, consider the customer experience provided by Dell in 1999. Dell's long waits and often-unhelpful phone support service for ordering and resolving technical issues frustrate Net users and corporate operation employees alike. The information provided on Dell's self-service Web site in 1999 was often agonizingly vague. Since the quickest way to get support is via the telephone, some Dell customers first check the self-service site and then call the support number to try and get more detail as to why their orders are late. "Double-dipping" with both online and phone support defeats the purpose of automating order tracking for self-service. A user-friendly customer service infrastructure is key to any value exchange that asks the Net user to perform an action.

Strategy Summary: User Support

Business Drivers: Community building, customer loyalty

Value Exchanges: Commerce, content, entertainment, promotion

Site Impact: Requires setting up channels for customer support as simple as an e-mail address or as complex as a self-service Web site for order tracking

Content Business Models: Retail commerce, pay per access, subscription, secure distribution

Best Examples: Amazon.com (*www.amazon.com*), TheStreet.com (*www.thestreet.com*), PETsMART (*www.petsmart.com*)

Communicate via Community

Beyond customer service lies a variety of communication techniques to build Net user relationships. A Net user community revolves around the content as a service. Net users visiting TheStreet's chat room are part of a community interested in financial marketplace news. Net users who write letters to the editor at Salon.com participate in a community by sharing their ideas. Even regular visitors to a niche content Web site, such as the popular garden enthusiasts destination Garden.com, are part of a community. Web communities consist of Net users with a common affiliation—the content on the Web site.

Customer service represents the foundation of communication with a Net user. At the minimum, the Net user must be able to track orders or contact a support representative if content or purchases are not quite as expected. But additional forms of communication build on that foundation to add more depth to the Net user's existing relationship with the Web site. Interacting with Net users can take the form of surveys, polls, message boards, and chat rooms. From a business perspective, it can be time- and resource-consuming to manage the flow of communication into channels such as polls and surveys. Message boards and chat rooms also require maintenance and infrastructure support. Generally, these alternate forms of communication are not effective without a sustained level of traffic on a Web site. It is difficult to maintain a useful and active message board without significant traffic (new or repeat) to post new copy. If a site has high traffic levels, then surveys, polls, message boards, and chat rooms add to the atmosphere of community. Let's consider how ESPN and TheStreet use different forms of customer interaction to build their Net user relationships on their highly trafficked sites.

Net User Communication Forums

ESPN SportsCenter (*www.espn.go.com*) uses polling as a mechanism to effectively drive repeat traffic and reenforce its position as a sports broadcaster in tune with fan concerns. Every day, ESPN asks Net users what they think about current issues in the sports world. Voting polls appear on the Web site asking for consumer opinions on everything from the college football player of the week to the most surprising rookie performance in the NBA. The Net user then selects from one of several answers provided. As soon as consumers vote, they are able to see how their opinions matched up with other participants'.

ESPN bolsters its position as a sports fan's destination by raising timely issues and providing interesting and timely sports issues for polling. Response rates help determine which issues are most popular with Net users. Polling represents one way that publishers can interact with users.

Discussion rooms and bulletin boards provide another opportunity to communicate with multiple Net users or to allow Net users to interact with each other. In 1999, TheStreet.com offered chat sessions on issues that tended to be the source of much discussion and debate. Visitors could check out the chat session transcript "after the fact" to read up on the expert answers provided to common questions. TheStreet.com hosted a tax chat with Tracy Byrnes from TheStreet.com and Martin Nissenbaum from Ernst & Young that fielded a wide variety of tax questions for Net users. Questions included general issues ("Is it really a good idea to do your taxes through the Internet?") and specific details ("I had a net loss in 1996 from trading. Can I go back and amend that return, do a 475 election, and then carry back to prior years?"). These types of discussions drive site traffic by providing a resource for interested Net users and reenforce TheStreet's reputation as a Net user resource center for investment activity.

Fantasy-based Communities

Fantasy-based communities, such as fantasy auctions and sports, provide a forum for Net users to interact around a common hobby. These communities provide a framework for users to demonstrate and expand on their expertise on a subject. They become traffic-generators for the publisher and also open up opportunities for syndication of related content and cross-selling of retail items related to the subject. Fantasy sports, a popular online activity, facilitate Net-user-to-Net-user communication using sports as the common ground. Fantasy sports tap into fans' loyalty and deep knowledge of sports and the players. Sites such as ESPN's enhance the community aspect of their Web sites by providing fantasy football, basketball, and baseball for their Net users. In 1999, Jupiter Communications reported that ESPN had registered more than 100,000 paying fantasy sports players. Net users can "buy" a team for $24.95 and start building a roster based on real professional ball club statistics (Keane, 1998).

Fantasy teams don't ever actually play each other. The competition is statistics-based. The Net user whose team has the highest stats wins the league. Fantasy sports tie together the online and offline worlds. For the fantasy basketball player, the activity links real-world basketball with a fantasy world where you are the coach, team owner, and general manager. ESPN supports the fantasy world by piping real events—such as player injuries, drafts, and trade information—directly to fantasy sports enthusiasts.

Likewise, a fantasy auction combines fictional commodities and real events. The Hollywood Stock Exchange® (HSX), launched in 1996, turned the fantasy auction idea into a fictional stock exchange involving the entertainment industry and quickly met with success (see Figure 5.5). HSX ties together real-world movie stars and recording artists with a fictional stock exchange, where Net users sell and buy stocks and bonds in movies, music, and stars. HSX has built a wildly enthusiastic community around its entertainment stock exchange. Thousands of users log in every day to check on stock "portfolios" that consist of "MovieStocks" (for actual films) and "StarBonds" (for anyone from blockbuster actors to newcomers).

The Hollywood Stock Exchange (*www.hsx.com*) makes Hollywood stars and their endeavors highly accessible to Net users by providing fresh content on movies, music, and actors to fantasy stock traders from all over the world. The entertainment-oriented stock exchange, with well over 400,000 registered users by 1999 actively buying and selling, is unique on the Web. Trading vol-

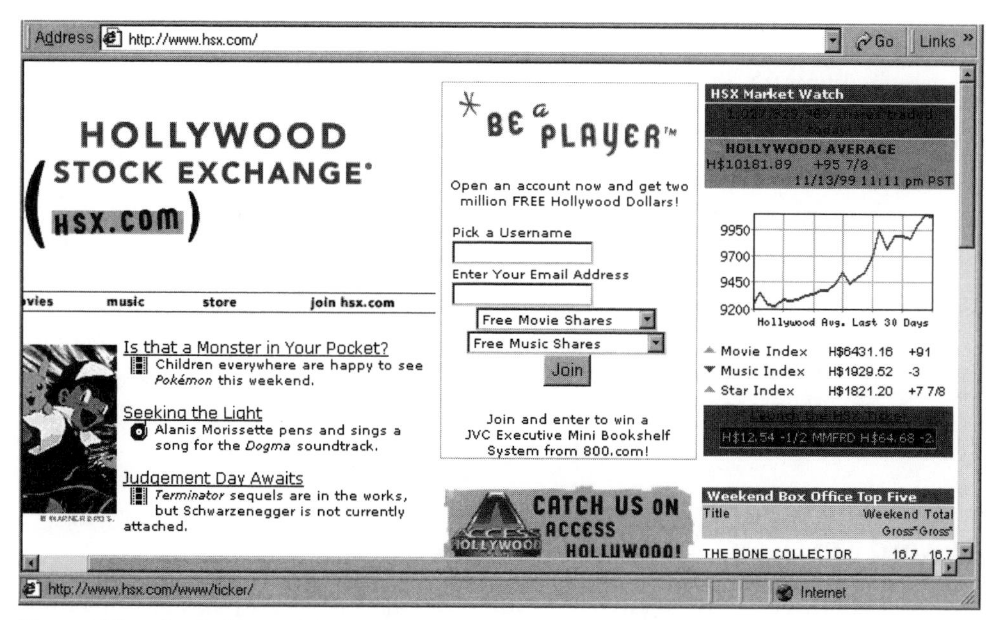

Figure 5.5 The Hollywood Stock Exchange.

Source: Used with permission of Hollywood Stock Exchange® (*www.hsx.com*). Pokémon™ is a registered trademark of Nintendo®.

ume averages more than 200 million "transactions" per day, and highly detailed trader fan sites proliferate on the Web. HSX makes its money by weaving real commerce links into the fantasy world. Net users reading up on the latest hot rock star stock can buy the artist's latest album from a music retailer by clicking on an affiliate link at the HSX site. Syndicated content from prime entertainment news sources currently includes *Variety* and SonicNet. (After all, you do have to do research on your investments.)

Building a successful community on fantasy auctions or sports sites requires a certain density of Net users. For smaller sites, this form of community is not a practical method for attracting loyal visitors. For larger Web publishers, though, the technique provides a nice touch of entertainment and innovation that forms a community around fantasy content.

Strategy Summary: Communicate via Community

Business Drivers: Community building, customer and site loyalty

Value Exchanges: Content, entertainment

Site Impact: Implementing a simple polling or survey architecture, while not really simple, is much easier than implementing auction infrastructure

Content Business Models: Advertising, consumer-driven commerce exchange

Best Examples: ESPN (*www.espn.go.com*), Hollywood Stock Exchange (*www.hsx.com*)

Reward Net Users

Publishers reward Net users for site loyalty in order to gain the following three benefits:

1. *Customer retention:* Net users return to the site because it provides them with benefits in addition to content, such as retail or content purchases.
2. *Increased frequency of purchases:* Net users increase the frequency of their purchases and, in some cases, the amounts they are willing to spend.
3. *Marketing for new customers and visitors:* Satisfied customers pass along the word in newsgroups and through e-mail about the Web sites that reward customers.

Publishers can offer Net users rewards for registering, making purchases, or participating in community activities. Award programs include sweepstakes, contests, loyalty programs like frequent flyer miles, and promotional currencies. Sweepstakes provide a traditional approach to engaging consumers in a relationship around content. *Salon* magazine offers contests that correlate with the literary focus of the Web site. One contest offers the opportunity to be published on the Web site and win $500 for writing the best "Brilliant Careers" essay. Another contest encourages readers to send in a seven-sentence pitch on why they should be chosen as the winner of a trip for two to the Independent Film Production Gotham Awards in New York. Salon. com is careful to link promotions with content. For example, to win a $250 gift certificate to your favorite restaurant, you have to prove that you are a "true Salon.com aficionado" by tying commentary quotes back to subject material.

Some loyalty programs offer *promotional currency*—nonlegal tender awarded to consumers for completing an action on a Web site, which can be redeemed for merchandise or other benefits.[3] In June 1999, InternetNews (Cox, 1999) reported that an NFO Interactive Survey of 1,093 consumers found that loyalty incentives played a significant role in bringing consumers back to a Web site to visit or even for repeat purchases (see Figure 5.6).

Site visitors and members benefit from rewards that tie into airline frequent flyer programs. Companies such as Netcentives (*www.netcentives.com*) provide a connection between airline loyalty programs and Web publishers through ClickMiles—a rewards currency redeemable for frequent flyer miles. Netcentives handles both the reward and redemption infrastructure for converting ClickMiles into frequent flyer miles. ClickMiles have a one-to-one relationship with frequent flyer miles from United Airlines, Continental, or a number of other major carriers. Consumers apply their ClickMiles toward free tickets or airline upgrades.

As with many programs that reward online Net users with offline benefits, actually putting your ClickMiles to practical use can be a complicated affair. Netcentives keeps an account for every Net user accumulating ClickMiles. Publishers pay the rewards service upfront for a lump sum of ClickMiles and then award ClickMiles to a Net user after completion of a defined action, such as registration. Once Net users reach a certain threshold of miles, they can redeem the ClickMiles in their account for real frequent flyer miles or continue to accumulate. For example, CDNow rewards consumers by offering one ClickMile for every dollar spent at the music store. Rewards are deposited weekly into the consumer's Netcentives account. In this case, CDNow rewards buyers for choosing the music store over other competitors who sell similar inventory at similar prices.

A British start-up called Beenz (*www.beenz.com*) offers a different type of redemption model for a promotional currency. Beenz is both the unit of promotional currency as well as the name of the company that manages awards

---

[3]For example, an Internet company called Flooz provides a gift certificate currency that can be used at a number of major retail sites.

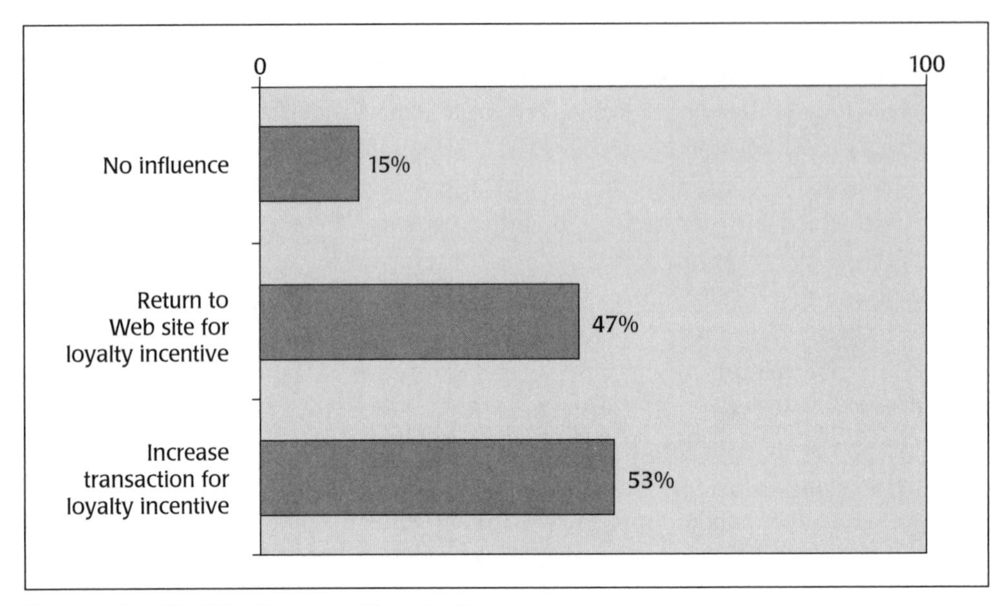

Figure 5.6 The Effectiveness of Loyalty Programs.
Source: From Cox, B., "Study: Consumers Want More Online Incentives." InternetNews.com, 28 June 1999. Used with permission.

and redemptions of the units. Merchants use Beenz to promote certain products and activities on their Web sites. Unlike ClickRewards, Beenz can be counted toward other purchases on the Web. The CountryBookShop (*www.countrybookshop.co.uk*), which offers more than 1 million British and American titles, awards Beenz to its customers for participating in the site's community. Customers get ten Beenz for completing five questions correctly on the site quiz, fifteen Beenz for adding a book-related question to the quiz, or twenty-five Beenz for writing a book review. Customers redeem Beenz at a small but growing number of Web sites (current list available at *www.beenz.com*).

While conceptually elegant, reward and promotional currencies can be difficult to implement, especially when tying an online action with an offline reward. Reward currencies frustrate Net users and publishers if technical glitches or user error prevents the Net user from getting the expected currency. Since the process straddles two sets of infrastructure (one for the merchant and one for the award program), backtracking through the award process in customer support can be painful. Another drawback to promotional currencies lies in the high threshold for redemption. For example, it takes thousands of ClickMiles to get a free ticket or upgrade. Meanwhile, smaller numbers of ClickMiles simply accumulate in a consumer's account and cannot be redeemed because they fall below the threshold. Beenz holders face the problem of finding merchants that accept Beenz as a payment mechanism. There are more merchants who grant Beenz as rewards than honor them.

In addition, publishers tend to award small quantities of Beenz. Merchants who redeem Beenz for content, products, and services ask for large quantities of the online currency. As a result, it can take a consumer a significant amount of time (and interaction with Web sites offering Beenz) to accumulate enough of the promotional currency to use. Customers in reward programs or with promotional currencies face the problem of accumulating multiple rewards while still falling short of redemption thresholds.

Both Beenz and ClickMiles give content publishers who do not offer commercial transactions a mechanism for rewarding consumers. Merchants who sell physical or digital goods have the option of using more traditional retail reward techniques that require the consumer to spend money. The award can be a percentage off or a free product in exchange for either frequency of purchases or total amount of purchases. Offering 10% off any purchase more than $100 provides incentive to get to the $100 threshold. The criterion for winning the reward is very straightforward: the customer must spend money. Whether the award consists of promotional currency or a commercial dis-

Strategy Summary: Reward Net Users

Business Drivers: Customer and site loyalty, promotional opportunities to cross-sell and up-sell products and services

Value Exchanges: Commerce, content, and entertainment

Site Impact: Requires an application server that offers built-in reward management, or outsourcing to a rewards program service such as ClickRewards or a promotional currency such as Beenz

Content Business Models: Pay per access, syndication, secure distribution, advertising

Best Examples: CountryBookShop (*www.countrybookshop.co.uk*), CDNow (*www.cdnow.com*), Garden.com (*www.garden.com*)

count, publishers can use reward programs to encourage repeat traffic and frequent transactions.

Market Effectively

Effective marketing of a Web site boils down to a clear communication strategy with potential and existing Net users. Marketing communication takes many forms, ranging from online (banners, buttons) to offline (television, magazine advertisements). Advertising in traditional media, such as print magazines, is often too expensive to consider for a long-term campaign. There are a number of lower-cost strategies that allow publishers to market their Web sites without venturing into bids for television spots. From a publisher's perspective, the benefit for marketing effectively is pretty clear: not only do you raise brand awareness and recognition, you gain traffic and drive purchases at your site. The trick is to understand and use marketing techniques that provide results without demolishing your budget. As a Web publisher, it's your responsibility to identify and use the marketing strategy that gives you the most impact for your budget. Seth Godin's book, *Permission Marketing: Turning Strangers into Friends and Friends into Customers* (Godin, 1999), provides an excellent in-depth study on effective marketing strategies for the Web. We will focus on providing a synopsis of marketing strategies in this section. For detail and case studies for effective implementation of the marketing strategies, see *Permission Marketing* or Seth Godin's other book *Emarketing.*

Advertising with Banners and Buttons

As the Internet grows more crowded, Web publishers realize that blindly inundating traffic with untargeted banners and buttons does not have a significant impact on customers. Forrester Research surveyed forty-seven Internet retail site marketing managers and asked them to rate the most effective advertising strategies from 1 (poor) to 5 (highly effective). Figure 5.7 shows that banners and buttons rank lowest in the experience of those surveyed, with direct e-mail and affiliate programs taking a clear advantage (Nail, 1999). The difference lies in the context provided with the advertising.

An untargeted banner or button is an advertisement that is presented to whoever views the Web page and happens to hit the advertisement's rotation.

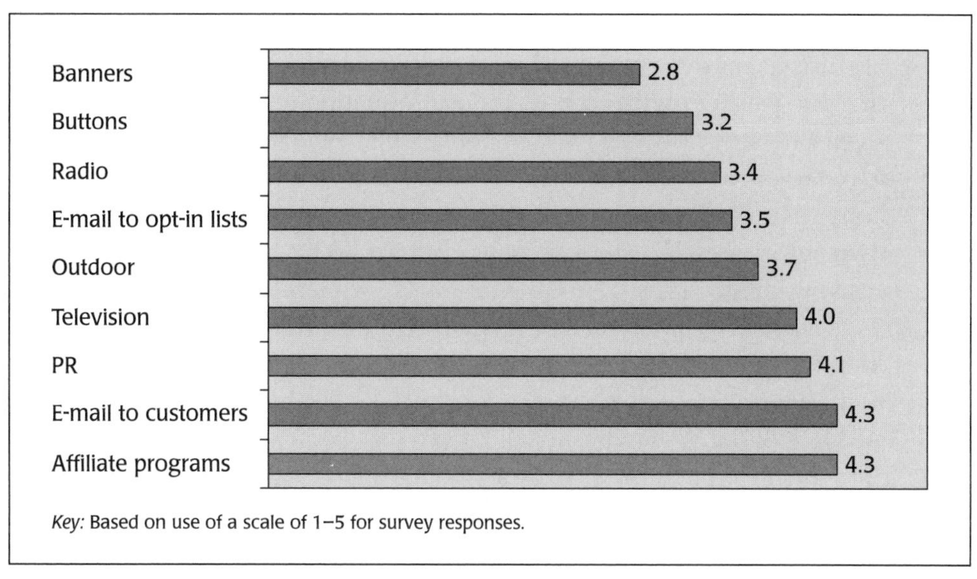

Figure 5.7 Effectiveness of Advertising Methods.
Source: Copyright © 2000, Forrester Research, Inc. Used with permission.

A targeted banner is an advertisement that is presented to a Net user based on the individual Net user's profile or the aggregated demographic for the Web site. Because of the predominance of untargeted advertisements, Net users are becoming inured to untargeted banner advertising content, which can have as low as a quarter of a percentage click-through rate. Banner ads are most effective when targeted at a customer profile[4] or a main demographic at a Web site. For example, Yahoo! Broadcast's main demographic consists of highly educated professional users with high-technology awareness. Consequently, Yahoo! Broadcast advertisers advertise luxury goods and electronic media sites, which results in a higher than 5% click-through rate on average. A growing number of online advertisement services are emphasizing the personalized ads.

DoubleClick (*www.doubleclick.com*), one of the most popular advertisement services, consolidates information about click-through rates for adver-

[4]A customer profile may indicate, for example, that the consumer falls into the category of those who have bought more than $50 worth of high-technology books within six months.

tisements from all sites in the DoubleClick network in a data warehouse. DoubleClick claims to serve more than 10 million advertisements a day across the Internet, which provides considerable information on ad click-through patterns. This service, which is called Boomerang, then targets different banners to Net users based on previous click-stream behavior stored in the data warehouse. The drawback to DoubleClick, as with other targeted banner advertisement services, is cost. DoubleClick charges both a high setup fee and high costs per CPM.

Two other promotional techniques are public relations and sponsorship. Both provide positive responses from Net users with little upfront cost. A positive blurb in a respected magazine such as *The Industry Standard* builds credibility for a Web publisher and encourages word-of-mouth advertising. Sponsorships consist of a sponsor (usually a large company) and a content provider that offers desirable content or services.[5] The relationship allows the sponsor to promote its brand alongside the content provider without the costs associated with content production. The content provider obtains relatively long-term (average sponsorship is three to six months) advertising revenue.

For the sponsored Web site, the motivation is simple: payment for the sponsorship via high CPM or a set fee. Sports broadcaster Quokka.com uses corporate sponsors such as Compaq to fund expeditions and the Web site coverage of expeditions. In return, Quokka hosts long-term banner placements for sponsors and offers editorial advertisements on its site.

Direct E-mail Marketing

If done properly, direct e-mail marketing can be the most effective marketing technique at the lowest cost. Direct e-mail marketing has between a 14% and a 22% response rate and often proliferates beyond the original e-mail list (Nail, 1999). Net users who forward interesting e-mail offers to friends and family extend the reach of the promotional campaign beyond the content publisher's efforts. In addition, commerce Web sites can customize e-mail lists using a Net user's purchase history or membership information. A digital image provider can send custom catalogs to Net users on record with purchases

[5]Sponsorships can be lucrative for content providers but can also result in significant administrative overhead for managing upfront and ongoing negotiation of terms with the sponsor.

for landscapes or wildlife, for example. A content publisher can make e-mails seem more direct by using first names, for example, when "talking" to Net users.

The trick to using direct marketing e-mail successfully is in a content publisher's willingness to respect Net user sensibilities about unsolicited e-mails. E-mail campaigns fall into two categories: prospecting and promoting. Prospecting campaigns tend to offend potential customers and should be avoided. Promotional e-mail campaigns provide existing customers incentive to visit a Web site and make purchases. Let's take a look at how prospective e-mail campaigns damage a publisher's credibility with potential customers.

Prospective E-mail Campaigning

The prospective e-mail route entails the purchase of general e-mail lists from an e-mail marketing company such as Net Perceptions (*www.netperceptions.com*). A content publisher or the e-mail marketing service then crafts e-mails describing offers from or the benefits of the site and sends everyone on the list the message. Prospective e-mails are *not* recommended for two reasons. First, prospective e-mail click-throughs have dismal return rates compared with e-mail campaigns with opt-in e-mails. ("Opt-in" refers to the consumer's willingly giving a company his or her e-mail address to be used by that company for promotional purposes.) Even if a content publisher buys targeted e-mail lists, which are considerably more expensive, it can't expect much more than a 2% click-through rate. Second, prospective e-mail is equivalent to cold calling and is detrimental to an attempt to foster a Net-user-friendly brand. You're bothering Net users in their private space (e-mail box) with offers that have little chance of interesting them. To the average Net user, prospective e-mails are junk and are commonly referred to as "spam."[6] In a 1999 survey on unsolicited e-mail by Jupiter Communications, a full 27% of Net users wanted unsolicited mail banned outright (jupstat, 1999).

Prospective e-mails have contributed to a rising Net user backlash against spam. Internet service providers, such as Mindspring and AOL, use anti-spam filters to block cross-posts to more than five news lists, or enact limits on how

[6]*Spam* (along with being an indestructible meat product) is loosely defined as unsolicited commercial e-mails sent to large numbers of consumers. Spam also refers to unrelated commercial postings to newsgroups.

many addresses can be placed on a "to" or "cc" line in an e-mail. Net users usually respond to unsolicited e-mails by deleting the message without looking at it or unsubscribing to any lists from which they did not want to receive e-mail. There's even a Coalition Against Unsolicited Commercial Email (CAUSE), which is lobbying Congress to enact legislation that sets guidelines for commercial e-mails and gives Internet service providers the authority to impose large fines for abusing the guidelines.

Any unsolicited e-mail sent to Net users has a high probability of being construed as spam. A Jupiter Communications survey of Net users found that the vast majority of browsers responded negatively to unsolicited e-mail and wanted it removed from their inboxes. The majority preferred unsolicited e-mail to be filtered out (either by service providers or by the e-mail client software) or banned outright (see Figure 5.8). The degradation in a content publisher's reputation is not worth the low click-through rates from an unsolicited prospecting e-mail campaign.

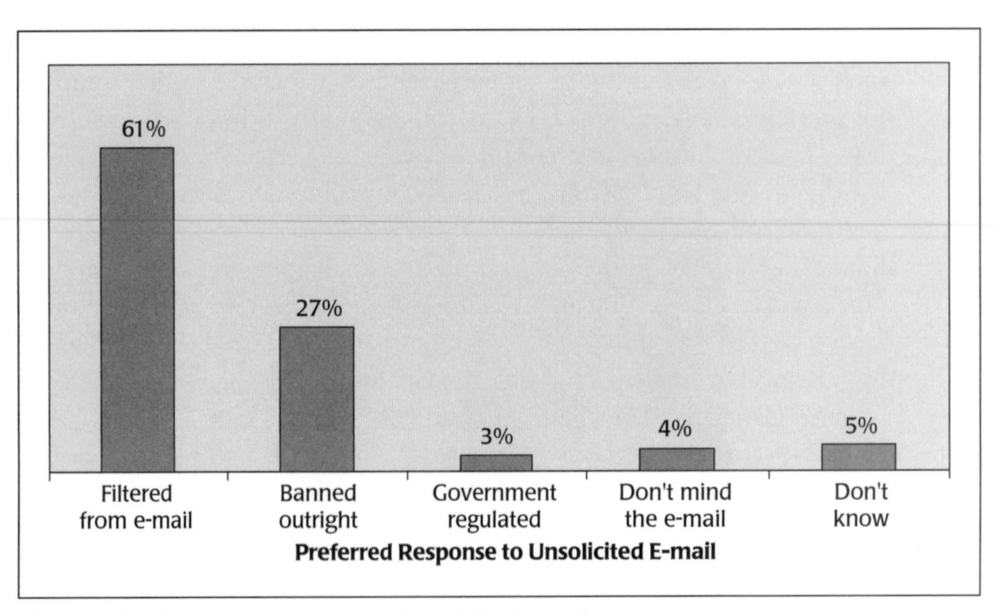

Figure 5.8 Consumer Response to Unsolicited E-mail.
Source: From "Unsolicited Email," Jupiter Communications Online Intelligence, New York: jupstat 2Q1999. Copyright © 1999 Jupiter Communications. Used with permission.

Promotional E-mail Marketing

The primary difference between a prospective and promotional e-mail is that a promotional e-mail is sent only to Net users who have specifically agreed to receive it. Promotional e-mail targets Net users directly while avoiding the risk of associating the brand with unsolicited commercial e-mail. A content publisher can embed links to special offers on syndicated content or pay-per-item products in the promotional e-mail, or can simply tell the Net user all about the latest and greatest content on the site. Effective promotional e-mails remind a Net user about the benefits of the value exchange and provide an incentive to return to the Web site for browsing or, preferably, a purchase.

Net users who indicate that they want promotional information have "opted in." Net users who decline the option of having content publishers send them promotional e-mails have "opted out." A Net user who has opted in has provided his or her e-mail address through a registration process on the publisher's site *and* has indicated that the publisher can use the address to send promotional information.

Depending on the content of the site, you can apply various levels to opt-in settings. For example, you can ask Net users for their preferences in promotional e-mails. Net users can opt in to promotional e-mails about financial content while opting out of offers on other types of content. Defining permissible e-mail content allows Net users to control the type of offers they receive and gives publishers the capability of tracking Net user preferences.

Prospective e-mail companies also sell opt-in e-mail lists (at much higher rates than blind promotional lists) that supposedly consist of people who have willingly agreed to have their e-mail addresses sold. However, it's better to err on the side of caution and not use these lists. A content publisher that does not have an opt-in list of Net users should start by setting up a membership base or partner with a registration service.

Other factors to consider are the frequency of the e-mail and the format. Even the most attractive offers get annoying if a Net user gets more than one a week from a single Web site. In addition, publishers should take advantage of most later-version e-mail applications' ability to render a message in HTML or rich text format (RTF), instead of plain text. The end result is much more visually attractive. The HTML e-mail message turns into a promotional Web page delivered to the Net user's inbox.[7] E-mail marketing companies report much higher click-through rates on HTML newsletters delivered to Net users than to simple text-based e-mails.

Establishing the infrastructure for e-mail promotions is an extremely complicated task. Companies face the options of building their own, buying an e-mail management solution from companies such as Kana Communications, or outsourcing individual campaigns to service bureaus such as Digital Impact. Many considerations factor into the decision, such as the extent of resources and infrastructure to manage the hardware to support e-mail list management and delivery. If you consider frequent and effective promotional e-mails a necessity for your company, obtaining an e-mail solution in house often proves the most practical solution in the long run. E-mail service bureaus, on the other hand, offer an alternative to the upfront costs of obtaining your own e-mail solution. These services take lists drawn up by publishers and define, execute, and manage e-mail campaigns. The cost structure generally consists of a one-time setup fee, optional consulting services, and a per-message fee (ranging from 3¢ to 5¢).

Promotional marketing can be a cost-effective and response-oriented way to engage Net users. Wine.com (*www.wine.com*), a wine retailer, uses Digital Impact (*www.digitalimpact.com*) to drive commerce and traffic to its site. Wine.com's e-mail list consists of well over 75,000 subscribers, many of whom are frequent purchasers at the wine retailer. Net users sign up for an e-mail newsletter by providing an address, which Wine.com promises not to share with other companies. E-mail is sent out every other week. The tone of each e-mail is familial, and the message includes interesting and often amusing discourse on the characteristics of a particular wine or the peculiarities of a specific year for bottling. Wine.com takes care to focus on the value exchange message in each e-mail. The newsletters provide important tactical information to Net users such as shipping deadlines close to a holiday and promotions on new items in the store. The overall result conveys a warm and welcoming invitation to use the Wine.com services as a resource for wine and fine living.

---

[7]Keep in mind that graphics and other nontext elements in an HTML e-mail are still linked to your Web server. An HTML e-mail crowded with fancy graphics and JavaScript will slow down your server performance if sent to a large e-mail list.

Strategy Summary: Market Effectively

Business Drivers: Customer and site loyalty, promotional offers

Value Exchanges: Content, commerce, advertising

Site Impact: Building out an independent e-mail infrastructure will have significant impact, although direct e-mail marketing can be outsourced

Content Business Models: Pay per access, syndication, secure distribution, advertising

Best Examples: Wine.com (*www.wine.com*), Quokka.com (*www.quokka.com*), Pets.com (*www.pets.com*)

Wine.com reports that sales generated from the e-mail newsletter generate more than 10% of the site's total sales (Williamson, 1999).

Set Up Smart Affiliate Relationships

An affiliate relationship allows one Web site to offer a link to another merchant or the merchant's product. Affiliate relationships let a Web publisher who does not actually own content or product market another merchant's products and earn revenue from the transaction. Even if the affiliate link is not based on a transaction, it can drive up traffic and brand recognition. From a publisher's perspective, affiliate relationships let pure-content sites collect revenue from transactions and traditional commerce sites widen the breadth of their Net user base. Affiliate programs are proliferating through the Net and are getting so popular that merchants are striving to outdo each other with affiliate rewards. Some affiliate programs, such as the one offered by Art.com, rewards referrers for every single purchase that is made by a customer who first visited the bookstore through the referrer's affiliate link.

Other affiliate programs, such as barnesandnoble.com, pay transaction commissions based on the total of purchases made in a shopping trip started from the affiliate link rather than the immediate purchase made after clicking the affiliate link. The Net user's benefit is equally compelling: an integrated marketplace of content and commerce based on Net user interests. To understand the nuances in affiliate relationships, let's talk about the two types of affiliate programs: portal-to-portal and site-to-site.

Portal-to-Portal Affiliate Relationships

Portal-to-portal partnerships involve a general portal or network and a vertical portal.[8] The general portal provides the benefit of a high volume of traffic and page impressions, while the vertical portal provides a service or product for the general portal Net users. The general portal gains much needed "stickiness" (the ability to retain users longer than the average 2.5 minutes per visitor) by providing an additional incentive (diverse content and services) for Net users to return. In addition, the general Web site gains a cut of the purchases made by Net users referred by the general portal. The vertical portal increases its transaction rate through the influx of traffic from the general portal. A general portal site earns anywhere from 4% to 30% of the revenue for the sale of a product promoted on its site (subject to the cost of the product and the demand for the site).

Site-to-Site Affiliate Programs

Site-to-site affiliate programs generally involve a pure-content and an e-commerce Web site. A publisher of a pure-content site registers with an affiliate program. The publisher then posts links to the commerce affiliate site. Every Net user who clicks through on the affiliate link and makes a purchase at the commerce site earns the content affiliate a cut of the revenue of the transaction. Affiliate programs award referring affiliates from 5% to 25% commission on sales. Site-to-site affiliate programs are generally sponsored by e-commerce sites or sites that mix content and commerce.

Signing up with an affiliate program is easy. To register, you simply provide your mailing address and site URL to the site's affiliate program and host

[8]Note that this type of program can also be merchant-to-merchant, if the two merchants sell different products.

the links that the affiliate program assigns you. The links contain a unique identification that lets the affiliate manager determine which referrals came from your site. The affiliate manager tracks referral fees as a percentage of sales or a small set fee (ranging from 25¢ to $1.50 per click for purchases generated from your affiliate links). Once your referral account reaches a certain threshold, the affiliate tallies the commission and cuts you a check. Net user publishers of homegrown Web sites can find affiliate programs to join using Refer-it (*www.referit.com*), a directory of affiliate programs.

Affiliation by Content

Affiliate agreements between professional content and commerce sites work best when the relationship is based on a common theme. Relationships that marry complementary pure-content sites with commerce sites work well because Net users obtain greater context around products before purchasing. Related content increases the click-through rates on affiliate links. If a Web site offers content that targets a vertical market, such as gardeners or auto enthusiasts, related content links provide added incentive to the Net user to click through. Examples of affiliate relationships are proliferating on the Web. Along with streaming media presentations, Yahoo! Broadcast offers affiliate links to Amazon.com to purchase videos or books. Entertainment-oriented Hollywood Stock Exchange hosts a variety of affiliate links, including some for music at the Virgin Megastore (*www.virginmega.com*) and *Star Wars* toys at Toys "R" Us (*www.toysrus.com*). Generally, affiliate links consist of text describing the link and a graphic of the commerce site's brand.

Some affiliate relationships use tighter integration. Take the affiliate relationship between Garden.com and *Horticulture* magazine, a popular print and online magazine for serious gardeners (*www.hortmag.com*). Garden.com is one of the premium Web sites for gardeners, offering a large selection of plants and garden ornaments for enthusiasts. Through Garden.com and *Horticulture*'s tightly integrated affiliate relationship, the *Horticulture Online* site shopping service is provided courtesy of Garden.com. Net users who follow the *Horticulture* Web site Garden Shop link end up at the Garden.com Web site (see Figure 5.9). *Horticulture* then takes a cut of any transactional activity that the visitor makes at the Garden.com Web site.

Tight affiliation between thematically related sites raises the risk of media cannibalization, which in this case means losing Net users or potential buyers

Figure 5.9 *Horticulture* Magazine Garden Shop by garden.com.
Souirce: Copyright © 2000. Permission granted courtesy of *Horticulture* magazine and *www.garden.com*.

by providing links to similar content on the other Web sites. When establishing an affiliate relationship between two related sites, it is important to minimize the Net user's exposure to similarities in content between the two sites. For example, Garden.com and *Horticulture* both offer their own magazines and chat rooms for serious gardeners, which means each site risks losing visitors to the other's content. When you are evaluating content affiliate relationships of your own, look for partners who offer goods or services that complement your own Web site without competing directly for purchases or (ideally) Web traffic. Well-thought-out affiliate relationships increase the value of content and products by offering complementary services for Net users while avoiding cannibilizing their member base.

Strategy Summary: Set Up Smart Affiliate Relationships

Business Drivers: Customer loyalty, new revenue streams

Value Exchanges: Commerce, content

Site Impact: Requires little effort to establish simple links, unless you set up a store integration such as *Horticulture* and Garden.com's

Content Business Models: Referral fee on pay per access, syndication, secure distribution, advertising

Best Examples: Barnes&Noble (*www.bn.com*), Garden.com (*www.garden.com*), Hollywood Stock Exchange (*www.hsx.com*)

Summary

The value exchange between a Web publisher and a site visitor depends to a large degree on the visitor's user experience. The experience results from the visitor's ability to interact with content effectively and enjoyably. The strategies suggested in this chapter focus on interaction, communication, and context. Publishers offer *interaction* through personalized Web pages, access to premium content, and forums for sharing opinions about content, such as fantasy gaming and polls. *Communication* plays a critical role in establishing a sustainable relationship with a site visitor. At a minimum, communication consists of a responsive mechanism for contacting the publisher. It can take diverse forms, such as direct promotional e-mails and marketing through targeted banners. *Context* allows a publisher to strengthen the relationship with the site visitor by enhancing the environment in which the content is presented. That includes providing Net users with a rich and multidimensional experience at the Web site. For example, affiliate links present the opportunity to visit other Web sites. Reward programs provide additional incentive to complete some desired action, such as registration, interaction in a site-sponsored activity, or a purchase. Each of the strategies described in this chapter help frame content in a value exchange. Publishers can implement each strategy individually or in parallel, depending on site objectives.

The user experience directly depends on the structure of the site. Consider the case of membership-based access to premium content. Offering premium content to members doesn't benefit the visitor unless the registration process itself is quick and simple. Likewise, any publisher planning on doing direct e-mail marketing should have a privacy policy, which is posted on the Web. Without these crucial elements of Web site structure, publishers can't fully take advantage of the strategies for killer content and service.

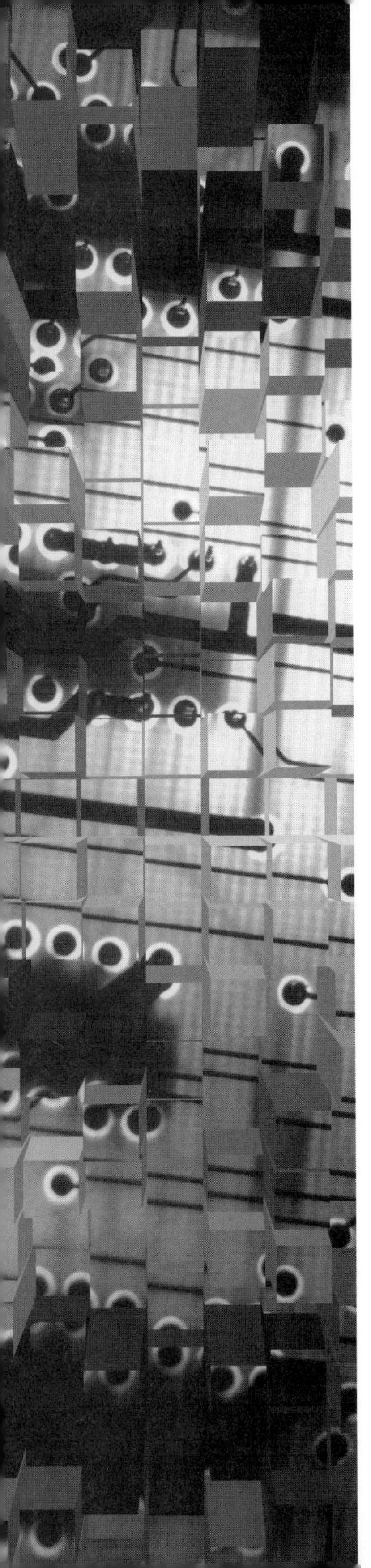

CHAPTER SIX

Designing Web Information Structure

A Web site's information structure and layout should support the consumer's ability to interact and use content and services. Without key elements (such as intuitive navigation) in place, a Web site visitor does not have the tools to enjoy and extend his or her site visit.

A poor information structure translates to an unappealing user experience. Lack of proper navigation features frustrates visitors and degrades usability. Even the most benefit-laden membership program can't succeed with a slow registration process. Table 6.1 lists the benefits of a properly designed site information structure.

Poor infrastructure equals lost opportunity for sales and advertising. Medium and large sites often struggle with poor infrastructure more than smaller sites. Publishers who manage large amounts of online information have to figure out how to manage its presentation effectively for consumers.

Consider the case of IBM's Web site (*www.ibm.com*). IBM pitches itself as an implementer and integrator of sophisticated, enterprise-level e-commerce systems. But until 1999, its own Web destination saw lackluster sales that were directly related to the poor content framework presented to Net users (Tedeschi, 1999). The **Search** option represented the most popular feature on the Web site

Table 6.1 Benefits of Good Site Information Structure Design

Structure Design	Content Infrastructure Benefit	Business Model Benefit
User-friendly navigation	Reduced end-user frustration	Increase in commerce, and advertising revenue
Targeted content	Increased probability of return visits	Increase in commerce, advertising, and promotion
Access content with minimal "friction"	Increased time spent for site visits	Increase in advertising

because Net users couldn't easily locate information in the muddled site architecture. (The **Help** button was the second most requested link.)

Poorly managed content creation and delivery contributed to the content infrastructure breakdown. The IBM Web site offered products for sale in three different sections: software, PCs, and peripherals. A separate product management and programming team owned each section. Each team, in turn, used many different design agencies (60 agencies in all). As a result of this mishmash, the Web site did not have a consistent look and feel across the more than one million pages. Search results differed by section. Pages loaded at inconsistent speeds. Potential customers who couldn't find the information they wanted simply left the site because IBM did not post a clear communication channel to customer representatives. IBM's Web site did not provide very encouraging proof of its e-commerce expertise for potential customers.

It took IBM ten weeks, more than 100 employees, and millions of dollars to rehaul the entire content framework, from internal content creation to infrastructure elements such as customer service. The redesign effort touched every single page on the site and included navigation, graphic design, and the arrangement of elements on the page. In the first week after the redesign, sales increased 400% and use of the **Help** and **Search** buttons declined dramatically. The site is noticeably fast, taking just four seconds to load over a modem.[1] The IBM "look" is standardized using templates for each page on the Web site. The interactive design agencies (which were pruned from sixty agencies to four) conform to a single coherent style using the site's standard templates. IBM's radical overhaul of its content framework demonstrates the importance of the relationship between site information structure and its content. In this chapter we will focus on other strategies like IBM's template standardization that support the consumer's interaction with content.

Table 6.2 describes suggested strategies to enhance a site's user experience. A screened sidebar summary of each strategy is included with the discussions in this chapter. The sidebars list key business drivers for the feature, the value exchange involved, the impact on the Web site, and the business model used. If applicable, the strategy summary also contains a list of Web sites or third-party vendors that follow this particular strategy noticeably well.

Use Consistent and Clear Navigation

Consistent and clear navigation depends on use of templates (or at least standard navigation bars) and a site search utility. Net users at a content publisher's Web site must be able to quickly and easily locate key categories of content and site information. Inconsistent placement of standard elements on pages on the Web site is disconcerting and sometimes confusing to users. In addition, any Web site that has more than three pages of content should offer intelligent search and/or indexing facilities. Net users come to a content publisher's Web site to find content or products. If Net users can't find what they want within a few mouse clicks, you risk losing that Net user, perhaps forever. Proper navigation facilitates the Net user's ability to interact with content and commerce on a Web page.

Web Page Templates

Web page templates enforce information infrastructure on a Web site. A template consists of a series of elements arranged in a set order on a Web page.

[1]An average Web page takes seven to twelve seconds to load over a 28.8 modem. A Web publisher should not post a Web page that takes longer than twenty seconds to load over a modem. Ardith Ibanez and Natalie Zee's book *HTML Artistry* provides practical advice for reducing the load time of an HTML page (Ibenez and Zee, 1998).

Table 6.2 Web Site Structure Enhancement Summary

Strategy	Description	Benefits
Use consistent and clear navigation	Provide a consistent design and navigation strategy across the Web site.	Customer and site loyalty
Support personalization	Categorize the content that formulates a personalization profile for the Net user.	Customer and site loyalty, promotional product upsell and cross-sell opportunities
Streamline form-based processes	Minimize the effort required to pay for products or to register with a Web site.	Commerce and membership programs
Practice open disclosure	Provide Net users with a privacy statement that clearly defines how you plan to use their information.	Site and member loyalty
Use appropriate digital rights management	Choose the most appropriate rights management method for your Web site.	Increased commerce

(Elements include standard tools, graphics, and text.) Templates can be set up manually during the content creation process or assembled on the fly by Web publishing systems such as Vignette's StoryServer and Allaire's Spectra.[2] Templates enable multiple members of a publishing community to maintain a consistent look and feel across pages. For example, the four agencies that work

[2]Templates, like any tool, can be abused. If your Web site templates consist primarily of static elements without enough room for new content, bored consumers will look elsewhere.

on the IBM Web sites depend on templates to ensure that updates and additions fit within a defined site design.

Site Search Utilities

Site searching and indexing remain fundamental requirements for any content site. Without the ability to quickly locate content, Net users are forced to click through multiple sections. Most Net users don't have the patience, especially if the desired content is buried in a larger article or section. Intrasite searching allows a site visitor to find information without browsing through the entire site. Search engines consist of three components: a "spider" that visits Web pages and follows links, an index that contains a copy of pages visited by the spider, and technology that matches entries in the index to the keywords in the consumer's search. Publishers can use metatags to add summary information for pages that may not have text to index, such as splash pages.

The *description* and *keywords* metatags provide information that the publisher wants the search engine spider to log in the index. The technology then ranks the relevant entries primarily by *location* and *frequency* of the keyword. If the keyword appears in a title, headline, or first portion of the text, the search engine assumes that the page is a good match with the consumer's search. Likewise, if the keyword appears in relationship to other words with frequency, the search engine assumes the page is relevant. Any further relevancy determination varies by search engine.[3]

There are a number of site-search utilities that provide intrasite searching, including AltaVista's Search Intranet, Microsoft Index Server, Netscape's Compass, and Infoseek's Ultraseek. Inktomi also provides a search engine that can handle more than 200 million indexed documents, which is used by high-traffic portals such as Yahoo! and GoTo.com. Select your site-search utility based on the size of your Web site, the importance of performance, and, of course, cost. Keep in mind that site-search utilities require maintenance. As pages are added and changed, publishers must rerun the site-search indexer, either on a scheduled or ad hoc basis. Most site-search utilities operate through incremental updates, adding only changes to the site index. Regardless of the tool you use, make site searching easy to find on your Web site and intuitive for Net users to use.

---

[3]Search Engine Watch (*www.searchenginewatch.com*) is an excellent resource site for searching techniques and infrastructure.

Strategy Summary: Simplified Navigation

Business Driver: Site usability

Value Exchanges: Commerce, content, entertainment, promotion

Site Impact: Consistent navigation metaphor affects all pages on the Web site. Site-search utility will automatically index a site, although publishers might want to add meta information about pages for more accurate indexing.

Content Business Models: Pay per access, subscription

Best Examples: Infoseek (*www.infoseek.com*), Excite (*www.excite.com*), IBM (*www.ibm.com*)

Support Personalization

Personalization of content builds Net user loyalty by customizing a Web experience. With syndication and affiliate programs, Web sites no longer contain content from a single publisher. Destinations compete to become "one-stop shopping" for content. Content publishers that began by specializing in a single type of content evolve into vertical portals that provide thematically linked content and services. Retail Web sites, such as Amazon.com, offer multiple stores, selling everything from books to records to electronics. To narrow down the flood of content, Net users use personalization as a form of self-service. By picking and choosing categories of content or services, Net users fashion their own unique experience at a Web site.

Personalizing Content

For personalized browsing, publishers identify general categories and group the appropriate content according to content attributes, such as author, subject, or price. For example, a music category for Jazz contains albums by John Coltrane, Wes Montgomery, and other jazz musicians. Publishers set up granular categories that subdivide general categories by elements such as time ("jazz musicians from 1930 to 1950") or type ("acid jazz").

Publishers can group content in a single category or in multiple categories, as appropriate. The publisher then allows the Net user to set up a profile that displays content or products based on these categories. From a customer's perspective, the reward is a single page of relevant, timely, and useful content. In return, the publisher gains a new member (if registration is the gateway to the profile service) and data on customer preferences. The publisher can also present targeted advertising and products based on the general categories chosen by the Net user.

Alternatively, publishers may personalize content and product offerings based on purchase patterns of consumers. This method gives publishers the ability to increase the rate and number of transactions for a content/commerce site. In order to offer purchase-based personalization, publishers require a site transaction history for each Net user and a means of identifying the Net user (ideally, at entry to the site).[4] Web publishing systems can offer targeted promotions or products based on previous purchases. This real-time marketing allows you to cross-sell, up-sell, and generally increase the rate of transactions at your site.[5] However, purchase-based promotions are truly effective only if you can identify the Net user at or near the point of entry to a Web site. Otherwise, you lose the opportunity to immediately engage a casual browser with relevant content and offers.

Using personalization means thoroughly understanding your own content, products, and their attributes. It also means offering an engaging and intuitive user interface for the personalization process. The Yahoo! personalization process, for example, takes just a few minutes and offers a simple user interface. In addition, your Web publishing system can't present personalized content or product offers in the right context if you do not do the work to catalog each content or product appropriately (by category, expected personal profile, or purchase pattern).

Minimizing the subjective nature of cataloging content improves the efficiency and appropriateness of personalized content. It's also important to develop a classification system for Net users as well as products and content.

Ultimately, your site publishing system will need to match Net users with content or product attributes. Companies such as Andromedia base their businesses on the science of making intelligent recommendations to visitors

[4]A common authentication practice is to store membership credentials, such as user name and password, in a cookie on the Net user's machine.

[5]Use discretion with targeted promotions. If a consumer is bombarded with a series of seemingly random offers upon entry to a site, he or she may become annoyed and leave the site.

based on Net user preferences and product attributes. One of Andromedia's customers, Cinemax (*www.cinemax.com*) offers a service called Movie Matchmaker that recommends movies to registered users. To get started with a personalization profile, Movie Matchmaker asks the Net user to rate twelve different movies. With this data, Movie Matchmaker builds a profile of the Net user's likes and dislikes. The more movies rated by the Net user, the more precise the profile.

Using Personalization Engines

Personalization engines provide the logic that maps content attributes with Net user attributes. To be effective, personalization must be in real time. An extremely robust system is required to offer real-time matches on Net user and product attributes. Personalization engines organize and sort content by attributes or other identifying information (metadata). The metadata about the content indicates its corresponding categorization to the personalization engine. The publisher then sets up business rules that map a personal profile or a general event (such as the first time a Net user visits a Web site) to a category. Finally, a personalization engine provides business reports on Net user personalization choices to help publishers fine-tune Web site content. In order to track the effectiveness of strategies, publishers need tools, such as Andromedia's Like Minds, that identify Web traffic and purchase increases and define clear preferences across the Net user base.

Catalog-centric e-commerce servers, such as Microsoft's Site Server Commerce Edition, provide basic support for personalization. Site Server includes a lightweight personalization server that lets you manage user access and profiles. Other publishing systems, such as Vignette's StoryServer, have much more robust support for personalization. For example, StoryServer handles adaptive navigation, personalization, and custom content delivery services, as well as storing visitor profile and behavior data. Vignette also offers built-in support for targeted content according to operating system and browser configuration. Allaire's Spectra specializes in offering tools that allow publishers to categorize content according to extremely specific personalization rules that apply to sections on a Web site or member profiles. For Web sites that just want personalization abilities without the other services in a content management system, Andromedia provides a (high-cost) scalable and effective personalization engine.

Strategy Summary: Personalizing Content

Business Drivers: Customer and visitor loyalty, repeat and targeted purchases

Value Exchanges: Commerce, content

Site Impact: Effective personalization requires that you carefully tag content and products by category

Content Business Models: Pay per access, syndication, secure distribution, advertising

Best Examples: Yahoo! (*www.yahoo.com*), Cinemax (*www.cinemax.com*)

Streamline Form-based Processes

Complicated payment and registration processes drive Net users away from the key sources of value on your Web site: purchases and membership. A content site with slow and complicated information collection will lose Net users to inept design. Net users abandon long forms except when highly motivated. From the publisher's perspective, adding registration and payment methods to a Web site comes with considerable overhead. Maintenance overhead includes membership databases, authentication processes, and access control to premium content. Collection payment also presents hurdles, such as real-time authorization checks on payment methods and fraud detection.

Retail catalog servers and Web publishing systems provide varying levels of support for collecting payment and registration information. One of the benefits of retail e-commerce servers is built-in support for payment infrastructure. While the payment itself is usually processed by a third party, a Web site must have the infrastructure to collect an order, pass the order details to the payment processing gateway, receive authorization, and provide a receipt or transaction history for the Net user. Merchant servers for e-commerce retailers package payment solutions as part of their product. However, these servers are structured around payment for physical products in a catalog rather than content. Therefore, once authorized, the purchase order is automatically routed through an order fulfillment process.

Selling content through a retailing e-commerce server is more complicated, although fulfillment is easier. Once a payment is processed, the Web site simply provides access to or a download of the purchased content. Some e-commerce retail servers allow programmers to build custom extensions that handle the sale of content. But the programming costs run high for both implementation and maintenance for these custom solutions. Content publishers have two additional choices: use a Web publishing system geared toward commerce or use an outsource service.

Registration presents similar design and infrastructure challenges. The two primary rules that govern registration are make it short and make it fast. Since registration generally provides access to a premium service or content, publishers don't need much more than a single unique identifier for each Net user (such as a name and password combination). The shortest registration process keeps to just those required fields, with e-mail an optional entry. Ask for more information only if you plan to use it. For example, getting address information lets you segment your member base by geography and perhaps try different content with different groups of users. Do not ask for information that can be construed as highly personal, such as a person's current marital status (unless it's expected, as in an online personal ad). Keep registration to a single screen if possible. Keep in mind that each extra page in the registration process substantially reduces the chance of the Net user completing the process.

Short and fast registrations are accomplished through short (preferably nonscrollable) forms and a clean layout. Some Web publishers use technology to make long and ungainly forms look presentable. Dynamic HTML (DHTML), which combines scripting logic and presentation tools, such as cascading style sheets, can make multistep processes fit on a single page without requiring a connection to the server for controls and graphics. The result is a faster experience along with a rich toolbox of graphic design and images. However, most publishers choose not to use DHTML for registration forms because DHTML is truly supported only by Netscape Navigator 5.0 and Microsoft Internet Explorer 5.0. In addition, since Net user actions, such as clicking a link, are processed by the Net user's browser, the publisher does not have access to any information about click stream data (that is, every page that the Net user passes through).

Accept Major Credit Cards

Credit card payment is the most popular form of payment on the Web today. According to a Forrester study ("Streamlining Web Payments"), 100% of US companies and 96% of European companies surveyed accept credit cards, making credit cards the leader by far in payment types (see Figure 6.1). Publishers use credit cards because they can authorize and settle transactions through a single financial institution. From the Net user's point of view, paying online with a credit card is a familiar and accepted process (Hagen, 1999).

If you have a significant European customer base, you may consider offering support for online debit cards, invoices, or Internet banking services for payment. However, be warned: these payment methods require integration work with third-party sources that may involve more time and effort than using a credit card payment gateway. If you can, avoid digital cash technologies and electronic checks, unless you have a large teenage or child customer base.

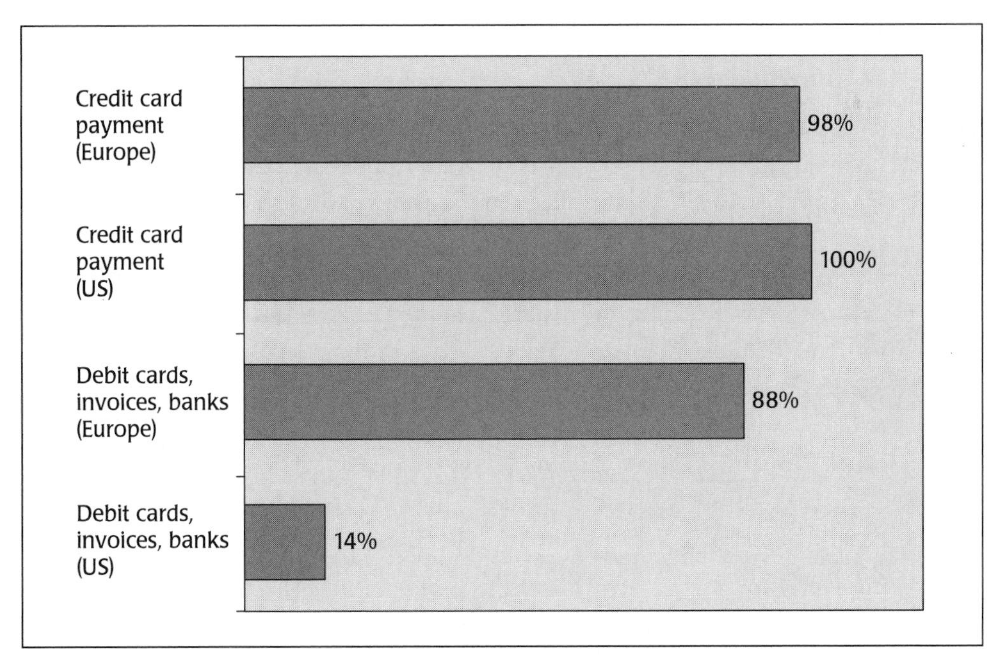

Figure 6.1 Most Popular Payment Mechanisms.
Source: Copyright © 2000, Forrester Research, Inc. Used with permission.

Handling Payments without a Credit Card

Less than 10% of teenagers have access to credit cards. Yet, Jupiter Communications forecasts that teenagers will account for $1.2 billion of e-commerce in 2002 (Jupiter, 7 June 1999). A number of vendors, such as RocketCash (*www.rocketcash.com*), DoughNet (*www.doughnet.com*), Qpass (*www.qpass.com*), and IcanBuy (*www.icanbuy.com*), offer accounts for teens and children. These "wallets" let parents transfer a weekly allowance payment to a Web account. The child or teenager then spends the money at any of the parent-approved sites. Parents are also able to restrict access to sites and even disable the account during certain times (such as during hours designated for homework). RocketCash has affiliations with popular shopping destinations such as MXG Online, Amazon.com, Beyond.com, and Pacific Sunwear.

Electronic checks are not popular with customers. They delay purchase fulfillment by up to seven days because using them requires waiting until the bank prints out the physical check and processes it. A seven-day delay is a liability if you're selling content and the Net user expects to get access to premium information in exchange for payment online.

Standardize Forms

At the very least, publishers should provide intuitive forms that Net users can fill in quickly and easily to register and submit payment. Right now, forms differ from site to site. Some sites require information that other sites do not, such as telephone number and company name. This means that the Net user has to learn a new payment process for every site visited. The major credit card associations and third-party payment providers are spearheading a movement to standardize forms. It was announced in June 1999 that Visa, MasterCard, American Express, America Online, Microsoft, IBM, Sun, Compaq, Cybercast, Trintech, and Transactor will back a digital wallet standard called Electronic Commerce Modeling Language (ECML), which is based on XML (eXtensible Markup Language). The first release of ECML defines the basic data required for payment collection as customer name, billing and shipping addresses, and credit card number and type.

XML in Web-based Information Exchange

XML provides a more structured and reliable way than HTML (Hypertext Markup Language) to transfer information over the Internet. As a language,

Strategy Summary: Streamline Form-based Processes

Business Drivers: Increased conversion from browser to buyer, increased transactions and registrations

Value Exchanges: Content, commerce, entertainment

Site Impact: Requires careful page design and fast performance for authenticating membership and payment

Content Business Models: Pay per access, secure distribution

Best Examples: Amazon's OneClick purchasing (*www.amazon.com*), Qpass (*www.qpass.com*), 1-800-Flowers (*www.1800flowers.com*)

HTML was not rigorously defined. Both browsers and search engines use considerable heuristics in interpreting a page. As a result, HTML is a poor language to use when transferring information that demands precision in parsing and presentation—such as e-commerce catalogs and documents. XML allows individuals to create Document Type Definitions (DTDs) that define the nature of every element in an XML file, which can then be sent over the HTTP Internet protocol. An XML parser uses the DTD to extract the right information from the file. Because of its structure, XML provides a platform to exchange information over the Internet more clearly and completely between databases or applications. ECML has not yet been accepted as a standard, but the guidelines it offers provide good advice when building your own payment collection forms.

Practice Open Disclosure

Net users' concern about privacy is increasing as more and more Web sites offer premium content in exchange for personal or payment information. Web site publishers do not have an effective means to gauge if their Net users worry about privacy issues. After all, a Net user uncomfortable with privacy issues simply leaves the site and doesn't return. But studies point to a growing worry among Net users that their personal information—name, e-mail address, even credit card data—is being sold to or shared between Web site publishers. Forrester Research analysts refer to this issue as the "privacy bomb."

A Forrester survey ("The Privacy Bomb"), shows that a whopping 61% of the Net users interviewed are "very concerned or extremely concerned" about privacy, and 80% believe that formal legislation is required to protect Net users' rights and privacy. The primary concern was not that information was shared but that it could be used by disreputable services. (In the real world, Net users divulge personal information in what they consider "appropriate" situations all the time. For example, you provide your medical insurance company with extensive personal information because you trust that the company will not share it with anyone other than your pharmacist and doctor.) Net users also fear that unregulated sharing of personal information increases the risk of being pestered by merchants at home or through e-mail (Delhagen et al., 1997).

Privacy Rights of Children

Of equal or greater concern are the rights of children, who may not have the discretion to judge what information should be provided to vendors. Unscrupulous vendors may ask a child to specify a parent's preferences or payment information. If, for example, a child is filling out a form that asks for contact information, the child may enter his or her home address—information that the parents have never agreed to providing.

Junk e-mail (*spam*), the telemarketing of the Internet, is the most common form of merchant abuse of Net users' e-mail addresses. If the Net user provides his or her e-mail address to a Web publisher and then starts receiving five junk e-mails a week, he or she attaches a negative association to providing information to publishers—even if the data results in targeted content or offers.

Posting Policy Statements

Individual commerce sites that do not make clear statements about how the site handles purchase-related information often exacerbate privacy concerns. Many commerce publishers attempt to collect payment data without providing even basic information about returns, refunds, or customer service. In 1999, the Federal Trade Commission conducted a survey that studied 200 Web sites, one half of them located in the US. Many of the sites studied lacked Net-user-related information such as refund policies and privacy statements. Only 26% posted a refund policy, 9% posted cancellation terms, and 10% posted any reference to applicable laws, such as export restrictions, that applied to the purchase. Perhaps most disappointing is that a full 35% of e-commerce sites surveyed did not post the total cost of the purchase (Masterson, 1999).[6] It's no wonder that many Net users lack confidence in handing over their credit cards on the Net. If your Web site offers registration, commerce, or personalization, you must post a privacy statement.

Privacy Statements

A privacy statement is essentially a contract with the Net user to protect personal information from misuse. A good privacy statement is clear about what constitutes proper usage of data. For example, the Earthlink (*www.earthlink.com*) privacy statement reads "We DO NOT sell our e-mail lists." The Billpoint Web site (*www.billpoint.com*) is particularly specific about its approach to children, saying: "This Web site is not directed towards children nor do we encourage children to use our services. Children are not eligible to use our services and we ask that minors do not submit any personal information to us." This policy provides an excellent example of clear and full disclosure of consumer information use.

Other Web sites provide advice for Net users, such as discouraging the use of social security numbers as site passwords. Perhaps the best solution is full disclosure of the publishers' intent in gathering the Net user's information. If you use data to target content and promotions, state the purpose of collecting information to point out the Net user benefit. If you plan to share Net user information with other publishers, provide a list of these publishers. Net users might not mind if the partner publisher has a respected brand and doesn't immediately flood their inboxes with promotional e-mails.

Provide an overview of how you use information and who in your organization has access to it. As with any contract, of course, you should not be boxing yourself into a legal dilemma. If you list content publishers you share information with, make sure the list is current. If a Net user finds out that an unidentified content publisher obtained his or her e-mail address from you, you could be liable for damages. Finally, make a distinction between individ-

[6]From InternetNews.com, 10 June 1999. Used with permission of Internet News/Reprint Services, St. Paul, MN.

ual data and aggregate data. Individual data provides detailed information on a single Net user. Sharing with publishers exactly where John Doe lives and what he bought last week at your site violates John Doe's privacy—unless your Web site clearly states that you plan to provide this detail to others. Aggregated information provides decision support without encountering the same conflicts with Net user privacy. For example, aggregated statistics (such as 50% of site visitors from New York City buy mutual fund reports twice a month) provide meaningful feedback on your Web site without intruding on the privacy of any individual.

Third-Party Seals

In addition to privacy policies, publishers use third-party organizations to "certify" that a site has acceptable privacy controls. Ideally, each individual approves the privacy strategy at a Web site before offering up personal information. Unfortunately, many privacy policies are couched in strict legal language so that Net users cannot understand the principles of the contracts. The World Wide Web Consortium (W3C) is in the process of setting up a standard for Web site privacy that automates Net user approval of a Web site privacy policy. The consortium is working on the Platform for Privacy Preferences Project (P3P), which aims to set a standard that allows Web sites to present privacy policies in a consistent way so that users can exercise preferences over these practices. The goal is to enable users to explicitly agree or disagree to a Web site's privacy practices. P3P benefits users by automating determination of a Web site's compliance with Net users' privacy expectations.

Net users can set up a user agent, which can be a browser, plug-in, or a proxy server, that looks at each Web site's privacy proposal and compares it to the privacy preferences set by the user. If the proposal maps to what the user wants, the agent accepts the Web site and records a "fingerprint" of it. If the proposal and user preferences don't mesh, the agent can inform the user that the Web site's privacy policy does not align with his or her wishes. The user then decides if he or she wants to proceed. Net users, therefore, do not have to read every privacy statement at every site. While a good idea in principle, P3P emphasizes user management of privacy preferences on custom software installed on the user's computer, which may prevent the standard from widespread or at least quick adoption by Net users. As a result, the onus falls on the Web site publisher to provide additional assurances.

Strategy Summary: Practice Open Disclosure

Business Drivers: Brand credibility, repeat site and customer activity

Value Exchanges: Content, commerce

Site Impact: Minimal; requires posting of text describing information usage

Content Business Models: Any that require personal information

Best Examples: Billpoint (*www.billpoint.com*), Earthlink (*www.earthlink.com*)

A privacy approval from a respected third-party site offers some measure of assurance for Net users that the site respects privacy issues. The two primary privacy certification associations are TRUSTe and BBBOnLine (the online version of the Better Business Bureau). Both associations say that to earn a privacy seal of approval, sites have to clearly notify visitors about their data collection and use practices and then get permission to use that data. TRUSTe provides a system of privacy seals (essentially, a Web site graphic) to inform Net users about the privacy practices on a Web site and also to provide a visual stamp of approval for a Web site's privacy policy. The BBBOnLine program consists of two seals: a general seal and another for businesses that advertise to children on the Web. Its framework calls for giving visitors "reasonable" access to their data, and it requires that sites take "reasonable" measures to protect the security of the information they gather. In addition, BBBOnLine prohibits collecting information on children under thirteen at children's Web sites without parental consent. Both BBBOnLine and TRUSTe have independent verification processes. BBBOnLine performs random audits, while TRUSTe conducts an annual audit of each member company.

Use Appropriate Digital Rights Management

Digital rights management focuses on preventing unauthorized distribution and usage of content. It breaks down into two main types: secure distribution and digital watermarking. Each solution provides a different type of user experience. Consumers of content distributed via secure distribution require software on their computers to access the secured content. Consumers of digitally watermarked content simply access the content through their Web browsers, without any need to download custom software. The publisher assumes the burden of monitoring usage using a special type of watermark-identifying spider.

Choosing the right solution depends on the value of the content and the market for the content. If you apply secure distribution to low-value content geared to consumers (such as an MP3 file with a single track), you risk alienating consumers who do not want to download the custom software required to access the content. But if you use digital watermarking for high-value content, you may be caught minimizing damages *after* someone had already distributed a $2,000 analyst report. In that case, prevention through secure distribution is more effective than tracking usage "after the fact" with watermarking. Let's examine the different tools and technologies available for each type of rights management solution.

Secure Distribution

Secure distribution depends on every potential user having custom software installed on his or her machine. It provides the most secure form of redistributing content by exerting business rules, such as pricing and promotions, on access to the content. Content providers with high-value content may select a secure distribution solution because the security of the distribution takes precedence over the hassle of the one-time custom software download. Consider the case of the $2,000 report. A Net user who purchases the report can redistribute the file via e-mail or diskette to any number of other Net users. The people who receive the illegally distributed report don't pay for the research and can, in turn, distribute it to other people. As a result, the content provider gets revenue from a single Net user when the product is acquired by any number of people. Along with lost revenue comes a more subtle dilution of the value of the content. The wider availability of the research reduces the value of the information and the incentive for others to pay full price for the report. Client-based digital rights solutions allow the file to be redistributed to multiple users but collects payment whenever a new consumer opens the file for the first time.

Here's how it works. Digital rights management (DRM) systems enforce a set of rules set up by publishers when packaging content for redistribution. For example, the publisher can set a rule that an analyst report in Adobe PDF format can be viewed unlimited times but only printed once after purchase. To access content, Net users must have software on their computers that can interpret the business rules. The software restricts access to the analyst report until the Net user fulfills a business requirement, such as payment or registration. After the Net user satisfies the condition, the custom software manages access to the report. While the Net user can view the report at will, he or she can't print the report more than once. The custom software acts as an electronic gatekeeper that checks the rules for content downloaded to the machine. If the Net user wants to access or use the content, the gatekeeper determines he or she has complied with the rules before permitting usage.

If the custom software on the user's computer is the gatekeeper, the publisher owns the mansion and defines the terms for opening the gate. The publisher dictates purchase business rules (price, promotions, and discounts) as well as access business rules. From a publisher's perspective, digital rights management allows you to sell content without worrying about piracy or intellectual property violations. Because you can manage how Net users access content on your Web site, you can encourage users to e-mail content downloads to their friends and family. E-mail becomes another channel for distribution. With a digital rights management system, you can set a rule requiring payment in order to decrypt and open a file. Properly used, DRM opens up your options for distributing content over the Internet to e-mail and alternate platforms such as set-top boxes.

As you might imagine, digital rights management systems are complicated and sophisticated. InterTrust, currently the leader in client-based digital rights management, has evolved an extensive technology platform using the Digibox—a new file format that packs up content, associated business rules, and, optionally, advertising associated with the content. Xerox, another company offering DRM, has spent years researching rights management issues at the Palo Alto Research Center (PARC). The end result is a full Web e-commerce system with digital rights management called Content Guard. This product includes tools for publishing, electronic commerce, user authorization, and usage tracking of digital documents. It also ensures copyright protection while maintaining document integrity and facilitating secure distribution, personalization, and a host of other services. But enterprise-level digital rights technologies can be difficult to implement and maintain in-house.

Licensing digital rights management technology is a solution only for extremely large organizations that have the flexibility to "humanize" the complicated technology. Digital rights system vendors provide the "guts" of the system—that is, the software infrastructure and tools that enable the rules relationships between Net user and content. Publishers or application service providers take the software tools and build user interfaces for Net users and/or publishers. In order to succeed with licensing technology, you need extensive system programming skills and expertise. Generally, only Internet service providers or application development houses have these resources.

Two options are to buy an entire e-commerce publishing service that's focused on DRM, such as Xerox's Content Guard, or to obtain a Web publishing system that provides support for rights management. The downfall to a DRM-centric e-commerce system lies in the possibility that the solution pays more attention to the management of content than to personalization and business intelligence needs.

Unless a publisher plans to make client-based digital rights management a core competency of the site, it's more practical to consider using a service bureau to outsource secure distribution. With outsourcing, you avoid disrupting your production environment while still managing rights to your content through the service's interface. PublishOne, a service built on top of InterTrust's DRM technology, provides a Web-based interface whereby publishers can package content into Digiboxes prior to delivery via the Web or e-mail—without requiring publishers to have the resources necessary to manage the complex rights management infrastructure.

Although client-based digital rights management places a burden on users to download custom software, it also allows content providers to distribute high-value information securely. For content that doesn't need that level of protection, there's digital watermarking.

Digital Watermarking

Digital watermarking works best for publishers who emphasize ease of distribution over access control. It places the responsibility for monitoring usage on the Web publisher and has little or no impact on the user experience. Digital image providers, such as Corbis (*www.corbis.com*), paved the way in the use of digital watermarking techniques to monitor and manage content usage. The music industry is also evaluating audio watermarking technology as a standard for distributing music on the Web.

Digital watermarks are a subtle way to identify ownership and origin of images. A digital watermark embedded in an image can be detected only by specific software, such as a watermark spider. Information stored in a digital watermark includes one or more of the following types of data: the origin, author, owner, distributor, and usage rules. The watermark may even include data that can trace the file even if it is processed, compressed, or distorted. Publishers embed the watermark in a file using a digital watermarking tool. One such tool, provided by Digimarc, is integrated with Adobe Photoshop so that graphic artists can attribute origin and ownership to individual files (see Figures 6.2 and 6.3).

Digimarc's watermarking tools come bundled with graphics applications from companies such as Adobe Systems, Corel, and Micrografx. Alternatively, graphic artists can obtain a Batch Embedder, which is used for large collections of digital images. For publishers, Digimarc provides a MarcSpider service

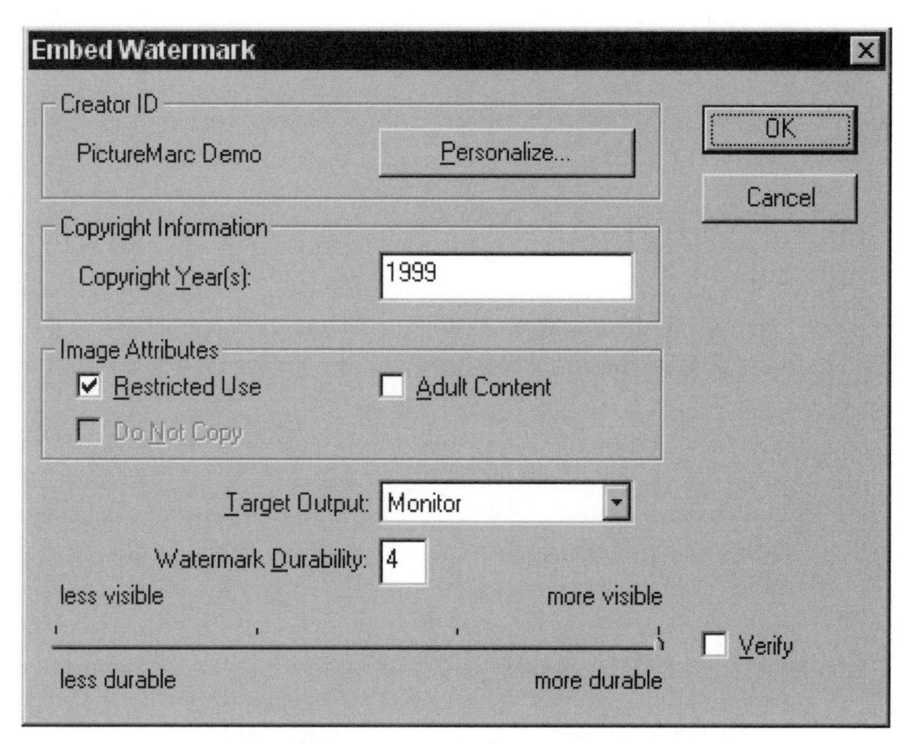

Figure 6.2 Embedding a Watermark through Adobe Photoshop 5.0.
Source: From Photoshop 5.0, copyright © 2000, Adobe Systems. Used with permission. Adobe and Photoshop are trademarks of Adobe Systems Incorporated, Seattle, WA.

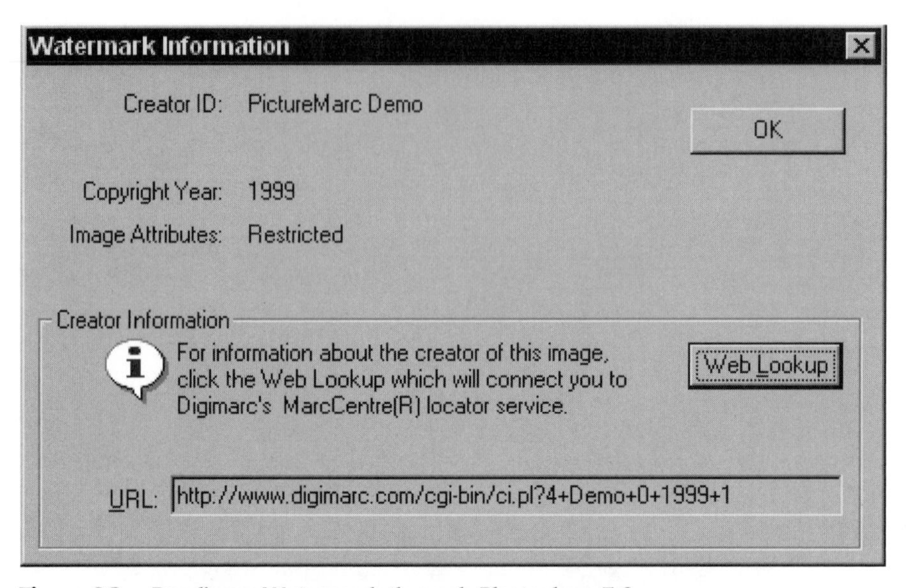

Figure 6.3 Reading a Watermark through Photoshop 5.0.
Source: From Photoshop 5.0, copyright © 2000, Adobe Systems. Used with permission. Adobe and Photoshop are trademarks of Adobe Systems Incorporated, Seattle, WA.

that searches the Internet looking for images with digital watermarking and reports on usage of the images.

The music industry trails far behind the digital image providers in adoption of watermarking technology. The five major record labels are taking preliminary steps toward investigating audio watermarking as a standard for secure distribution of music. In August 1999, the Secure Digital Music Initiative (SDMI) announced that the committee planned to adopt ARIS Technologies' digital audio watermarking system as part of an early effort to secure portable digital music players. ARIS digital audio watermarks are embedded within audio signals, which can be music files, multimedia files, or audiovisual works. ARIS watermarks contain data such as program content description, program sources, and program recipients. But it's too early to commit to a technology for the secure distribution of music because the music industry still remains conflicted about the nature of music distribution on the Web. As a result, publishers of audio files risk losing the investment in a technology if they select a rights management technique too early. Digital image watermarking is far more promising for publishers of medium-value content.

Strategy Summary: Use Appropriate Digital Rights Management

Business Driver: Ability to distribute and track high- and medium-value content

Value Exchanges: Content, entertainment

Site Impact: Bringing rights management solutions in-house requires resources and ongoing maintenance. Alternatively, some companies outsource rights management.

Content Business Models: Pay per access, syndication, subscription

Best Examples: Corbis (*www.corbis.com*), Art.com (*www.art.com*)

Summary

A Web site's information structure plays a vital role in supporting the user experience and value exchange. This chapter has focused on the *information organization* element of site structure. Information organization describes the presentation of content on a Web site. Strategies include clear and consistent navigation, streamlined registration processes, and the appropriate level of security in content distribution. The ability to effectively personalize content contributes to a positive user experience. Information organization also applies to how a publisher uses visitor data. A site's privacy statement defines the ways a publisher categorizes (and uses) a visitor's personal information.

What about the availability and applicability of tools and technologies to support the implementation of these strategies? The final chapter of this book focuses on the tools and processes that enable killer content. It's impossible to discuss every solution that supports value exchange, so we will focus on the key elements. Understanding these internal processes contributes to a publisher's ability to build or enhance the right production environment.

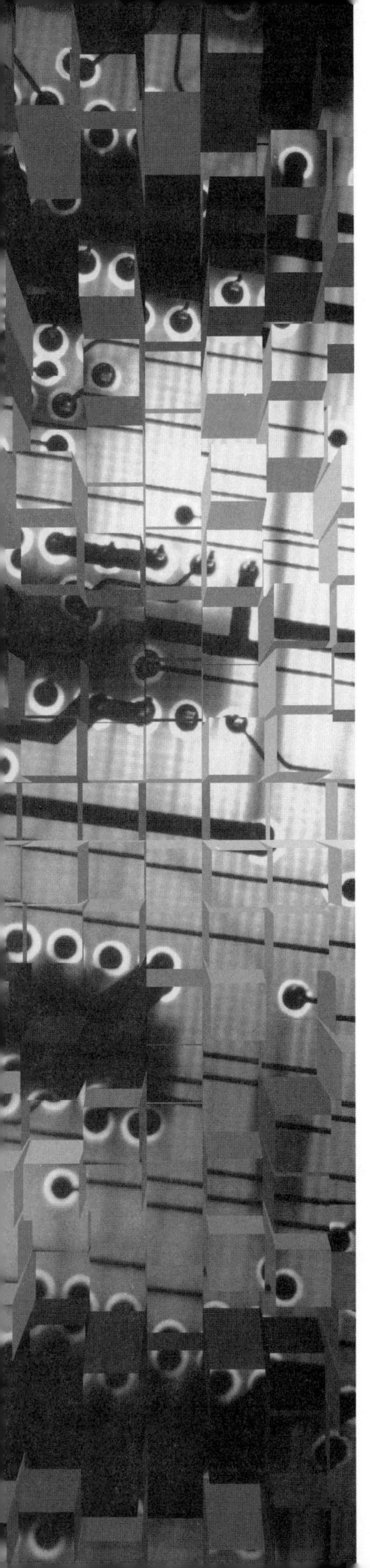

CHAPTER SEVEN

Facilitating Web Site Processes

A publisher's production environment consists of computers, software, applications, and technical and marketing contributors. This chapter focuses on the *workflows* that support the processes around value exchange on a Web site. Understanding the benefits of different workflows helps frame how to manage killer content and also select the right tool or technology for a publisher's individual environment. An effective process infrastructure enables publishing organizations to create, publish, and manage frequent content updates, as well as to perform more sophisticated analysis of content browsing and purchase patterns. Site infrastructure and processes support the following production goals:

- Maximize productivity
- Minimize site maintenance costs
- Reduce time to market for new content and updates
- Proactively manage hardware and software capacity in the production environment

Elements of a Process Infrastructure

The elements of a process infrastructure play a critical role in the support of the user interface and Web site structure (see Table 7.1). Information infrastructure, as described in

Table 7.1 Elements of Site Process Infrastructure

Name	Description	Examples
Web server*	One or more servers that process HTTP requests for Web pages and applications on the site	Microsoft Internet Information Server, Apache Web Server, and Netscape Enterprise Server
Content management server/storage *(optional)*	Stores, identifies, and processes elements used for assembling Web pages; some content management servers, such as Vignette's StoryServer, have built-in content management processes that automate production workflow. Other servers, such as Allaire's Spectra, contain in-depth analysis tools for analyzing the performance and capacity of the Web site.	Vignette StoryServer, Allaire Spectra, Open Market
Application server *(optional)*	Enables the execution of Web-based applications and programs; an application server can be geared toward streaming media formats, as provided by Liquid Audio, can focus on object-driven applications, such as Allaire's Cold Fusion, or can provide multiple, lightweight site management solutions such as Microsoft's Site Server.	Liquid Audio Liquid Server, Allaire Cold Fusion, Microsoft Site Server
People and processes	The business and development functions and personnel involved with producing and managing a Web site	Marketing, development, Web administration

*For simplicity's sake, network architectural details, such as Web farms or integration with payment processing components, are not included.

chapter 6, focuses on the presentation and substance of content and services in the user experience. Process infrastructure supports that user experience through tools, third-party relationships, and production workflow. The efficiency and organization of the production environment make a significant difference in a publisher's ability to develop and maintain killer content. Depending on the size and complexity of the Web site, a production environment may contain multiple elements.

Efficient collaboration between the different elements of the publishing environment improves efficiency of processes, reduces the time to Web site for new content, and shrinks opportunity for user errors. Publishers of large Web sites and/or those whose content cycles are frequently updated must have a publishing system that manages the workflow of the production environment (see Figure 7.1).

Vignette provides an environment geared at high-volume, content-based sites. Erik Josowitz, vice president of product design at Vignette, explains that StoryServer customers, such as CNET, ZDNET, CBS Sportsline, and Snap!, use the content management system to drive sites that experience as many as 10 million page views per day. In addition, Vignette focuses on supporting production workflows with hundreds of participants. For example, the *Chicago Tribune*'s Digital Community used Vignette's StoryServer to manage the content workflow of "a significant portion of the authors and editors who work on the standard print and online editions, plus a new cadre of authors and editors focused only on community news." The workflow involves almost one thousand editors and authors making updates to online newspaper content more than once a day. As Josowitz puts it, "In the publishing space the notion of release has given way to continuous publishing" (E. Josowitz, phone interview, 20 Aug. 1999). Continuous publishing of content that changes at a high frequency demands a high-performance system. When asked about StoryServer's ability to scale, Josowitz points to existing customers as examples. CBS Sportsline uses StoryServer to manage content production on their popular sports Web site. Sportline sees page loads rise to more than 15 million page views per day during the Final Four college basketball tournaments. Checkout.com, an e-commerce site that depends on personalization, is directly marketed to more than 8 million people in the Los Angeles area alone. A system, such as StoryServer, lets a Web publisher manage and support high-volume Web sites that can be maintained with frequent content updates.

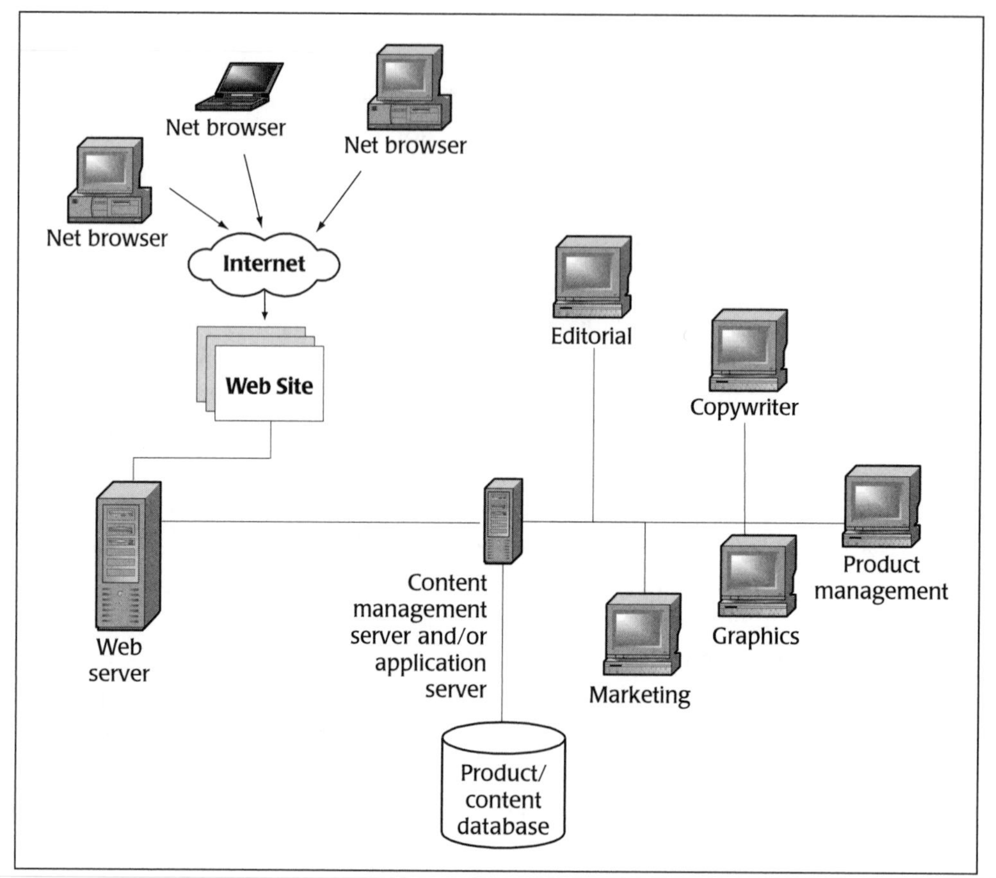

Figure 7.1 Publishing Environment Example.

Allaire Systems offers a lower-cost solution for Web publishers who have less rigorous requirements for workflow management and content updates. The Spectra system continues Allaire's tradition of flexibility and extensibility by giving publishers the tools to easily extend the site management user interface and processes to individual environments. In addition, Allaire layers Spectra on top of its Cold Fusion application server, which provides a rich environment for Web-based applications and utilities. Spectra emphasizes different things from StoryServer, such as role-based security (for membership databases, user directories, access control, and permissions models) and highly detailed business decision support and technical monitoring.

Jeremy Allaire, founder and Chief Technology Officer of Allaire Systems, sees Spectra as one element of an Allaire platform, providing tools, development environment, and delivery solutions specifically developed for the Internet. According to Allaire, Spectra is "a higher-end system intended for companies who know what they are doing and know what they want to do." The solution provides an open architecture and extensive application program (called a Content Object Application Programming Interface, COAPI) geared toward intermediate and advanced Cold Fusion developers. Spectra provides content management, business intelligence, system management, personalization, customer management through e-mail, and a host of other publishing features. The software even integrates and bundles RealNetworks' streaming media server (RealServer) to support rich media types. Allaire points out that one of the key benefits to the Spectra component and the Allaire platform in general is that the "content process infrastructure is directly integrated into the enterprise Web application infrastructure," complete with a generic, mainstream application server and popular visual tools (J. Allaire, e-mail interview, 3 May 1999).

While most publishing systems aim to solve similar problems in workflow management, different solutions place emphasis on different areas. Vignette, with its suite of StoryServer, Syndication Server, and MultiChannel Server, looks to capture the high-end market for high-volume Web sites that are updated by a number of participants. Allaire's Spectra system fits in nicely with the other Allaire tools (Homesite and Cold Fusion) to present a lower-cost and highly extensible architecture for content management and presentation. When evaluating a content management system, consider the size of your Web site, frequency of updates, number of participants in workflow, development resources, and cost in the equation. A high-volume content portal differs in priority of requirements (and resources) from content providers with less volume. To help illustrate those differences, the remainder of this chapter describes strategies for enhancing process infrastructure. By better understanding how process infrastructure provides business benefits, you will be able to evaluate and enhance a content management system.

Table 7.2 Strategies for Enhancing Process Infrastructure

Strategy	Description	Benefits
Understand Content and E-commerce Solutions	Understand how publishing and e-commerce systems solve different problems for consumers	More informed tools evaluation
Manage Process Relationships Effectively	Automate manual processes in the production workflow	Quicker content updates, fewer errors in delivery
Take advantage of Business Intelligence	Leverage the results from browsing and purchasing patterns to fine-tune navigation and merchandizing strategies	Commerce from better targeting of offers, proactive planning for when to add hardware to production environments
Develop Scalable Affiliate and Syndication Management Programs	Affiliate relationships are profitable only when they can be managed in an automated way	Commerce for content sites, greater brand recognition

Table 7.2 summarizes suggested strategies for enhancing the processes behind a well-maintained Web site.

Understand Content and E-commerce Solutions

Pure-content Web sites have different priorities and even different requirements from e-commerce sites. Publishers pick their systems based on the systems' ability to support value exchange on the site, whether that value exchange is for retail goods or for content. Most content management systems have limited support for e-commerce fulfillment—billing, collection, and shipment infrastructure. Likewise, e-commerce merchant servers, such as Microsoft Commerce Server and Open Market Transact, provide many features that simply are not as relevant to content publishing as they are to hard goods e-commerce. Order fulfillment is a key example. Order fulfillment for online retail sites is, in a word, complicated. There's inventory to verify, shipping and payment information to collect, orders to send down a supply distribution chain, and customer exchanges and refunds to process.

Delivering content is much more simple. Content is delivered online through the Internet or offline through e-mail. If you're a content publisher, the extensive support for order fulfillment in a merchant server goes unused. On the other hand, content publishing systems generally have limited support for handling e-commerce order processing, but can interface with third-party components or build-your-own solutions to handle inventory and billing for small inventories.

Previous chapters introduced the notion of complex processes that support the user's interaction with content, such as secure distribution of content or personalization. To some Web publishers, these features do not warrant the resources for implementation and maintenance. Similarly, other publishers may not be concerned with access control if they do not require registration to view premium content. Each Web publisher prioritizes his or her requirements according to the value exchange for the Web site. Once that evaluation is complete, it's time to pick a system that works for the publisher's individual production environment. Identifying categories of environments will help distinguish between e-commerce and publishing system solutions.

Catalog Solutions

Traditional e-commerce environments are *catalog-centric;* that is, the production process infrastructure handles the supply, demand, and fulfillment of products in an online catalog. Many of the Net retailers base their Web presence on inventory. Catalog-centric solutions work well for stores that sell products that are tied to a backend order fulfillment process. For Web sites that mix content and commerce, catalog-centric environments provide little benefit and require considerable customization to use effectively (see Figure 7.2).

The average retailer using a catalog-centric solution cares more about volume of goods than community on its Web site.[1] Simple rules govern content on the catalog-centric Web site. If the product is available in inventory, the

[1]Sophisticated retailers like Amazon.com do offer some community-building features, such as message boards for book reviews.

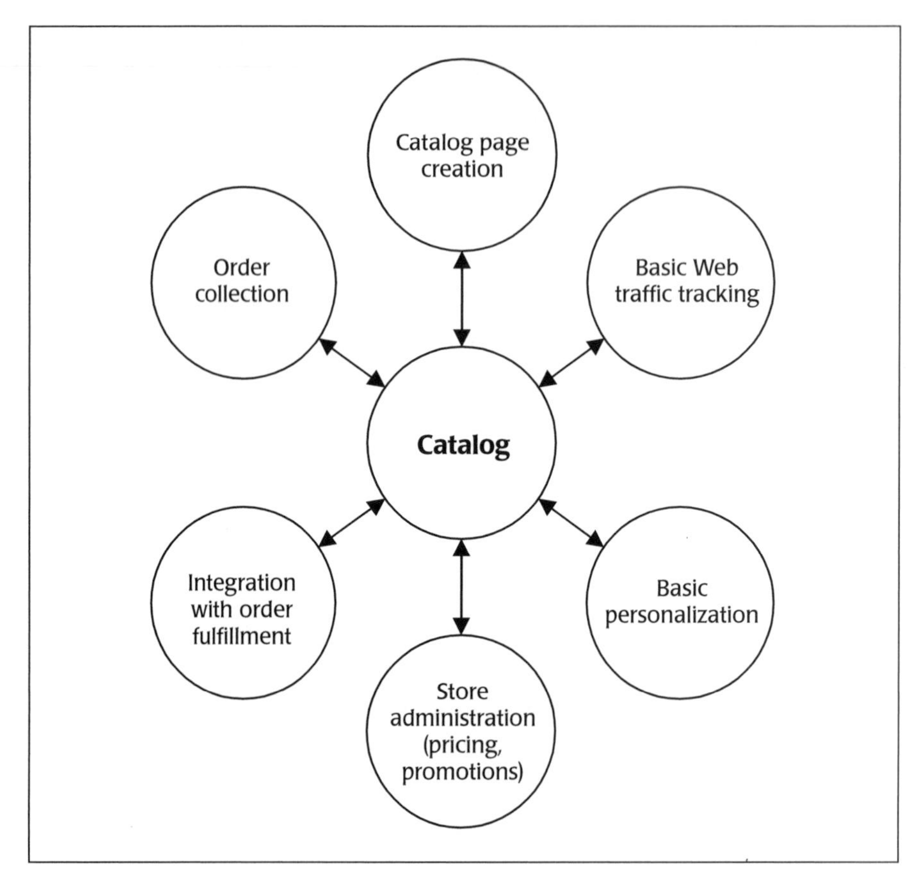

Figure 7.2 Catalog as a Focal Point in Retail E-commerce.

Web site displays the product and its associated attributes. If the product is not available, the Web site displays a message indicating that the product is out of inventory and, potentially, puts the product on backorder. A merchant server such as Microsoft Site Server Commerce Edition provides excellent support for inventory management but has limited flexibility in managing dynamic content, other than inventory-oriented promotional offers.

Building more sophisticated personalization and workflow improvements on top of a catalog-centric merchant server requires extensive custom development or outsourcing. Companies such as PeopleSoft solve workflow management problems but at a cost many mid-size Web site publishers can't support. Building custom extensions or rolling an entirely new solution is not an attractive option for many content publishers who do not have, or want to add, programming resources.

Catalog-centric solutions do have their benefits for publishers who emphasize commerce over content on their sites. Servers such as Microsoft's Commerce Server offer store templates and a store-building wizard, which can get an online retail store up and running very quickly.[2] The Commerce Server extensible Order Process Pipeline allows merchants to extend the basic e-commerce system to include integration with accounting and other third-party products. But content publishers who are looking for a maintainable and flexible publishing system for content-driven sites turn to process- or content-oriented solutions.

Process Solutions

Process-centric servers focus on the workflow processes that drive a content commerce Web site. Process-oriented publishing systems care most about productivity in the publishing environment. A process-centric publishing system such as Vignette's StoryServer knits together the different functions that produce a Web site, as well as provides limited commerce functionality. Each participant in the content management cycle focuses on a single core competency and uses the publishing system as the basis for collaboration. Copywriters work in Microsoft Word to provide the proper wording for content. Graphic designers use tools like Macromedia to build images and animation. Programmers code HTML and server-side scripts in their favorite integrated development environment (IDE). Each member of the production environment uses publishing system templates to route new content for approval. Workflow productivity improves because none of the participants in the content development process work outside of their core competency. Workflow processes drive the content into key elements in the publishing environment (see Figure 7.3).

Process-centric solutions closely resemble real-world publishing, which is one reason why application servers such as Vignette's StoryServer are so pop-

[2]Microsoft's BizTalk provides an XML framework for companies to communicate files, content, and other services over the Internet. BizTalk's XML standard appears to be the only planned support for workflow processes by Microsoft.

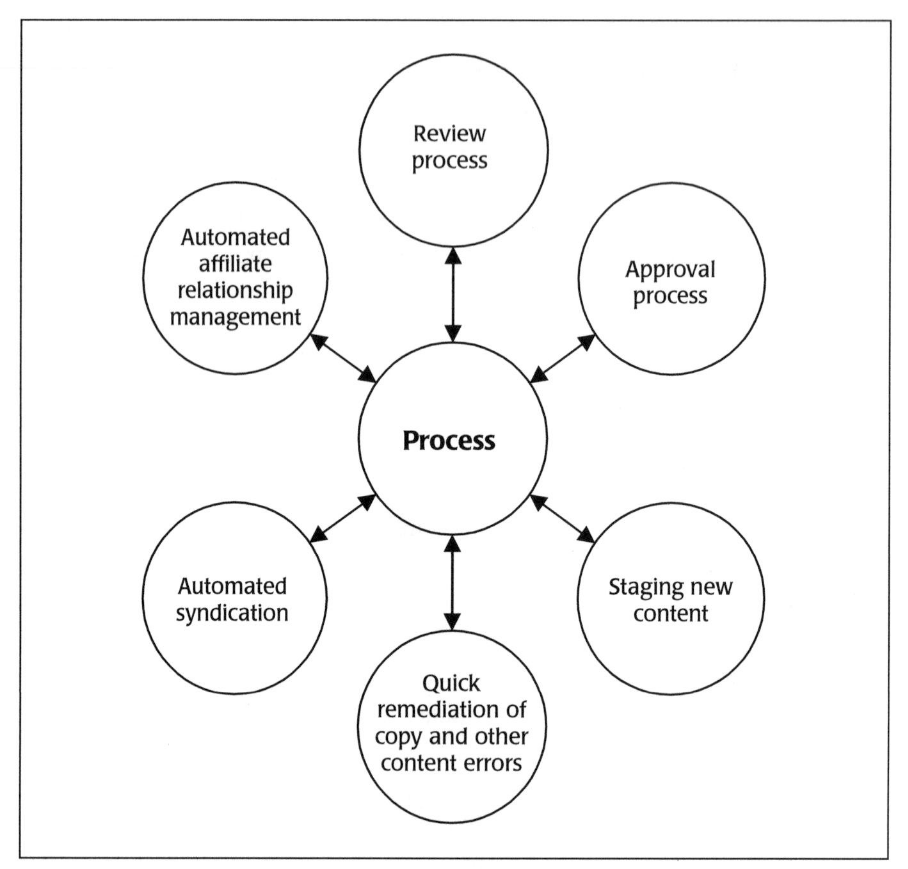

Figure 7.3 Process as a Focal Point in Publishing Environment.

ular with online magazine and newspaper publishers. The approval and change control for content in a Vignette workflow template is very similar to the process of approving a print journalist's article for a newspaper. In addition, Vignette provides a publishing system that business managers use just as easily as programmers and marketing personnel.

Application Solutions

While process-centric solutions do provide personalization and syndication services for more ambitious Web sites, publishers with an eye toward maxi-

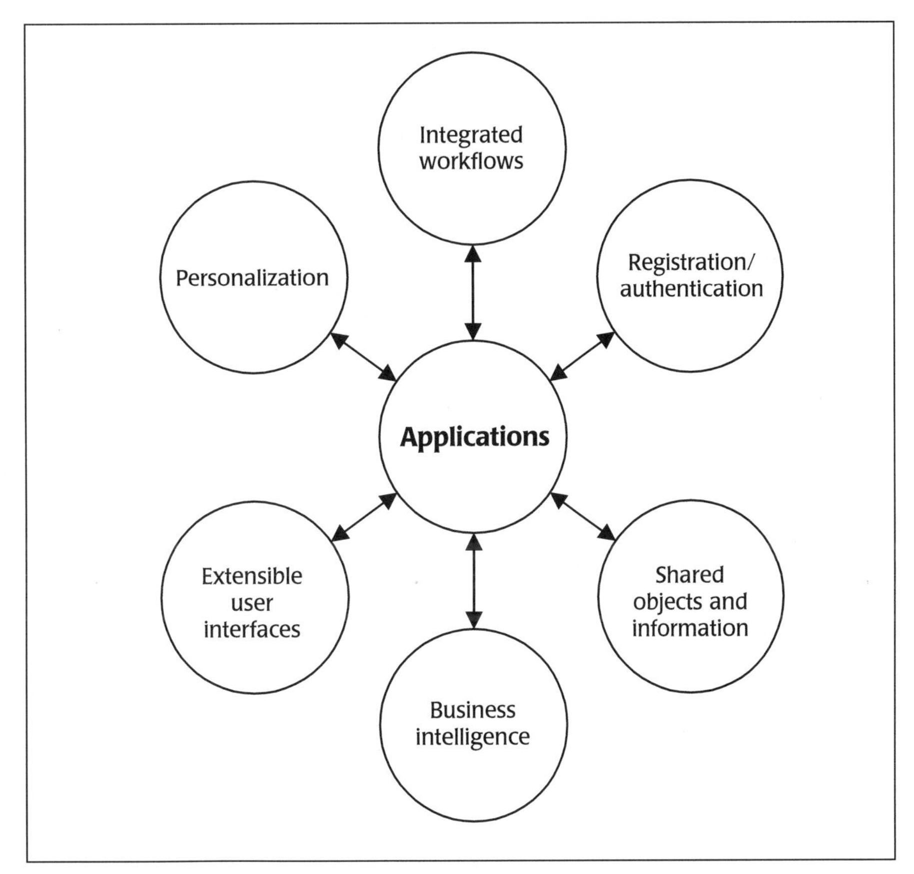

Figure 7.4 Applications as a Focal Point for Publishing Environment.

mizing the dynamic nature of their sites turn to application-centric solutions (see Figure 7.4). *Application-centric* solutions drive content according to variables such as Net user preferences and syndicated information. Application-centric servers provide publishers with the greatest level of control over what gets displayed on the site, whether it's through affiliate partnerships or through personalization. Applications include personalization engines, payment engines, registration processes, and other processes that automate or enhance services at a Web site.

Customization drives workflow processes as well as the storage and presentment of content. Application-centric solutions are ideal for publishers of

Strategy Summary: Understand Content and E-commerce Solutions

Business Driver: Better ability to make a tools evaluation for a production environment

Value Exchanges: Content, commerce, entertainment, promotion

Site Impact: Selecting the right type of publishing-commerce system significantly impacts the content updates and processes that support a site's value exchange

Content Business Model: Any

Best Examples: Microsoft Site Server Commerce Edition, Vignette servers (StoryServer, Syndication Server, MultiChannel Server), Allaire's Spectra

sites with the resources to take advantage of the flexibility in developing custom content, processes, and workflows. Application-centric solutions, such as Allaire's Spectra, give Web publishers programmatic and user interface flexibility and control over workflow processes, personalization features, and a host of other services. Spectra also provides extensive support for order processing and other transactional activity.

Manage Process Relationships Effectively

Net users value content for its relevance, timeliness, and utility. Web sites that host outdated ("stale") content risk losing visitors to more current destinations. To raise productivity and reduce the time to post new content on the Web, a publisher must establish effective process relationships for the organization. The first step to managing production processes effectively is to admit that a publisher is never "done" with updates, enhancements, and other changes to content. Content is subject to inventory changes (such as new products or services), internal changes (such as a new press release), and external factors (such as daily reports from the field during an international crisis). Each of these changes churns through the content management cycle.

Automation of most of the manual processes improves the speed and responsiveness of site updates.

Content Management Cycle

The content management cycle involves people with a variety of different skills, ranging from the business professional (journalist or analyst) to the programmer (HTML or server-side) to the design and production team (graphic artist, Web site producer). Other participants in the process include the product manager and finance accountant, both of whom are highly interested in the Net user response to Web content. Each member of the team collaborates throughout the content management cycle (see Figure 7.5).

Content *creation* (building graphics, Web pages, and Internet applications) often requires programming as well as design expertise. Content *management* refers to the oversight of a wide range of workflow processes, ranging from approval of inventory to posting of content on the live site. Producers

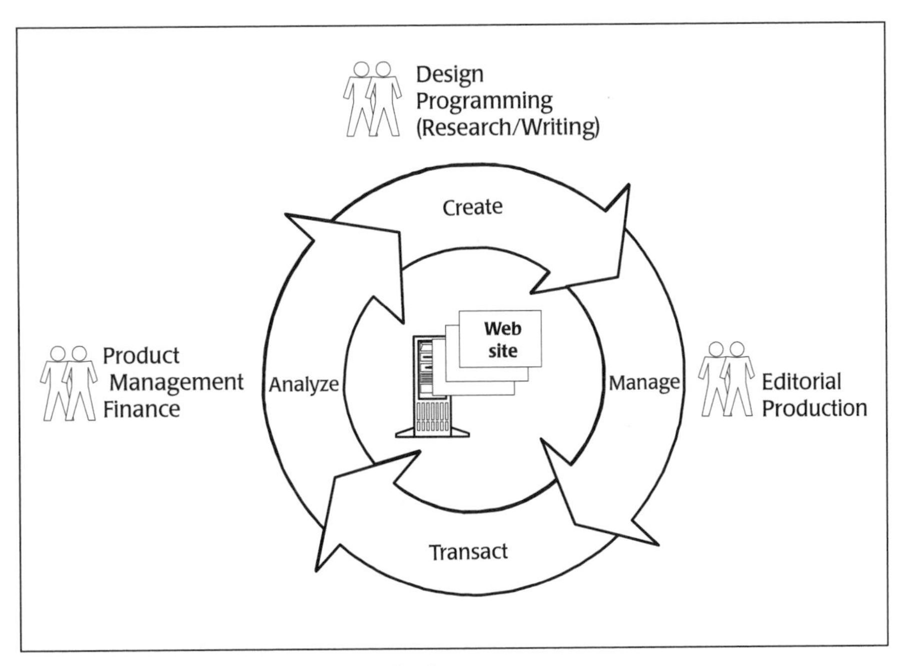

Figure 7.5 Content Management Cycle.

and editors manage content from its inception to its location on a live site. Servers that collect orders and process *transactions* for Net users are also needed. Finally, product managers require tools to *analyze* traffic and purchase patterns in order to fine-tune product development and business relationship strategies. Accounting is also interested in the net revenues and purchase patterns. Depending on the complexity of the process and the tools used, the interval between content creation and business analysis can be as short as one day or as long as several months. "Co-location," or the hosting of the Web servers in a separate facility, can further complicate the process, since in that case local business processes must be integrated with the remote operational ones. Automating any of these processes in the content management cycle, through scripts or through a content management system, reduces human error and time-to-market.

Automating Workflows

A highly productive and efficient production workflow reduces the overall cost of maintaining a Web system. It maximizes the productivity of all the contributors. In addition, a well-designed process is able to grow with the resources and Web site infrastructure. Without automation of workflow processes, you risk bogging down contents and services in the mire of production mechanics. One of the primary benefits of a Web publishing system lies in its ability to set up workflow templates for a production environment. A workflow template defines the process that an element undergoes en route to a Web site. That process includes the creation, approval, categorization, and presentment of the content or service. Figure 7.6 demonstrates how a workflow takes a photograph from a physical image to a graphic integrated on a Web site.

Many different parts of the organization participate in the process, including the business (marketing and copywriters), development (application engineers and Web developers), and IT (Web site administrators) departments. To maximize productivity, design a workflow that ensures each participant operates within his or her own core competency. Copywriters' contribution to the production environment is text. Writers generally use an office productivity tool, such as Microsoft Word, to generate content. Asking the copywriter to write marketing copy in HTML reduces the writer's ability to produce timely quality copy. Separating the content from the form allows the different partic-

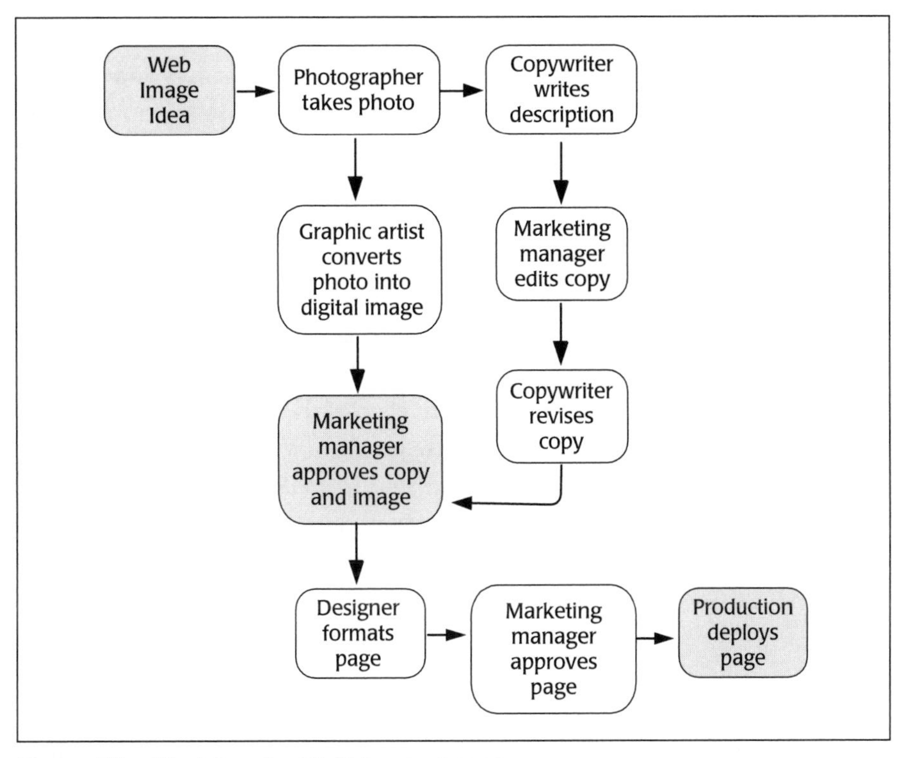

Figure 7.6 Workflow for Digitizing an Image.

ipants in the workflow to operate in a comfortable environment while working together on a product.

Web publishing systems provide a range of services to facilitate workflow management, including design tools for the layout of the site, a model for indexing and searching content that is in production or under development, and a method for profiling content for personalization. For example, Vignette's StoryServer, which specializes in workflow management, offers well-designed templates that map to standard content management workflows. Allaire's Spectra provides similar automated workflow management through a Web-based interface. For example, with Spectra a business manager can use the Web interface to initiate a task and then assign the task to another participant in the environment. To better understand how to build efficient workflow processes, let's take a look at the requirements for each of the participants in the production environment.

Business Users

Business users include marketing managers, business managers, product managers, copywriters, editors, and producers. A workflow can revolve around a single change such as a press release, or a new service such as enhanced search functionality. Business managers must define and initiate content changes and then approve them. Other business users need a common interface to define items to be published, assign publishing schedules to these items, and make selections for personalization rules and settings.

Business users require intuitive and familiar tools to participate effectively in content production. Luckily, Microsoft Office 2000 provides strong support for saving Word documents relatively cleanly in HTML. Microsoft 2000 uses XML to preserve formatting between Word and Excel files and Web servers, which results in more reliable transfers from document to the Web.

Business users can create, approve, and work in a familiar tool such as Word and then convert the document into HTML. If the Web site uses page templates, it's a simple process to transfer the HTML code to a standard site template or even have the Word document converted directly to the template.

Creative Services

Creative services receive assignments from business users and contribute design elements to a Web page under development. Creative departments include graphic artists and media programmers who work with streaming media forms such as video and audio. Creative services users require a central interface for receiving assignments and providing content for approval.

One of the primary goals for the workflow should be to allow participants to use the most effective content creation tool. Whereas business users employ easy-to-use office productivity tools, creative services employees use complex tools that require expertise and training. For example, graphic artists often use Adobe products, such as Photoshop or Illustrator, or Macromedia, to build images and alter photos. Media programmers use rich media tools such as Apple's Quicktime 4, Microsoft Windows Media Toolbox, RealNetworks' RealProducer, or Macromedia's G2 to develop streaming media and video for Web sites. Just as you want to preserve the creative service users' ability to build content effectively, you also want your workflow to shelter other users from working with unfamiliar graphic tools and file formats.

> **Strategy Summary: Manage Process Relationships Effectively**
>
> **Business Drivers:** Quicker time-to-market, less opportunity for manual error
>
> **Value Exchanges:** Content, commerce, promotion, entertainment
>
> **Site Impact:** Impacts all areas of a Web site, especially those that need frequent updates
>
> **Content Business Models:** Subscription, syndication
>
> **Best Examples:** Vignette's suite of servers (StoryServer, Syndication Server, MultiChannel Server) and Allaire's Spectra

Developers

Developers generally fall into two categories: those who build Web-based applications and those who build Web sites. Developers use the workflow to receive requirements for projects from business users and hand off the results for approval.

Take Advantage of Business Intelligence

Business intelligence reporting gives you the tools to understand the effectiveness of the content value exchange with site visitors. *Business analysis* reports on Web site usage based on site traffic patterns and trends, including any transactional information and user response to marketing programs. *Production monitoring* reports on the usage of servers supporting Web sites and site applications, including real-time monitoring and analysis of historical trends for operational usage.

Business performance indicators for your Web site include information from Web server logs, such as the number of page hits and the referring Web location for the visitor. Other information that can be derived from the Web logs includes length of visit and user path through the site. This type of information feeds back into the design process of a site's structure. If Web server logs show that the Help section of a Web site is the most-often-used page, that points to navigation problems on the site. Web traffic statistics are essential to understanding how best to present and target content to site visitors. Publishers either build their own systems for reporting on Web traffic or use a third-party product such as WebTrends, Andromedia, or Marketwave.

If a publisher offers retail or affiliate purchases, there's even more valuable data to be mined. From transaction information, you can derive data about the frequency of purchases per customer, rate of purchase per product or content, and other data that tracks commerce value exchange on a Web site. From a publisher's perspective, commerce business intelligence gives you the tools to determine how you're doing today against your Web site mission and how to forecast and plan for tomorrow. If a registration database is maintained, the publisher can analyze site activity by demographics and other site visitor characteristics. Reporting interfaces for transaction information include online application processing (OLAP) tools and report writers such as Seagate Crystal Reports.

An often-overlooked area of business intelligence is in server performance and capacity. Production monitoring helps publishers understand when the capacity of the site databases and servers affect the customer's user experience. Tracking that operational data historically and forecasting trends over time inform a publisher's production growth plan—such as when to add new Web servers before the load from Web traffic and purchases degrades a Net user's visit.[3] Real-time monitoring satisfies different but equally important objectives. It allows you to respond quickly to issues in your production environment, such as a failed server. For example, BMC Patrol, a network monitoring service, offers a consolidated console for servers in a network. If a server goes down, BMC can send an alert to a pager or a voice mail system, so that an IT group can respond quickly. Historical and real-time monitoring provide critical reporting for the network components of your site process infrastructure. Hosting service Exodus acquired the Colorado start-up Service Metrics specifically to monitor the performance and activity of the Exodus environments. If your Net users can't access parts or all of your Web site with a certain degree of

[3]If severely overloaded, a Web server may crash and take down a database that stores purchase or member information.

Strategy Summary: Take Advantage of Business Intelligence

Business Driver: Ability to fine-tune purchasing and navigation strategy based on reports

Value Exchanges: Content, commerce, entertainment, promotion

Site Impact: Requires tracking of business intelligence information through Web server logs by members and by purchase transactions

Content Business Models: Advertising, pay per access, syndication, subscription, secure distribution

Best Examples: Exodus, Seagate Crystal Reports, BMC Patrol, WebTrends, Andromedia, Marketwave

performance, you risk damaging their confidence in your Web site's ability to provide content and service reliably. Business intelligence provides the tools for publishers to understand how to improve presentation on and reliability of their Web sites.

Develop Scalable Affiliate and Syndication Management Programs

Affiliate and syndication programs allow publishers to put commerce and content where the site visitor demands them. Affiliate links let content publishers promote a retail merchant's product on their Web site in exchange for a cut of the end transaction. Syndication subscriptions give a single Web site the ability to provide content from a variety of different content partners. Both affiliate programs and syndication benefit customers by providing content and commerce within their areas of interest. For example, I'm a big fan of Gary Payton of the Seattle Sonics basketball team. I check out the Sonics home page (*www.seattlesonics.com*) to see who's injured, upcoming game schedules, and other information. But if I want to get the latest analysis of how the team is doing this season, I have to leave the Sonics Web site and go to ESPN.go.com or some other sports site for commentary. If the Seattle Sonics used syndication, the Sonics home page might have commentary and analysis syndicated from ESPN.go.com, NBA.com, or the Sporting News.

Consider the experience from the merchant's perspective. A fan arrives at the Web site and is immediately presented with a variety of information about the team and individual players. The leap to purchasing a game ticket is considerably less than if the Web site has less engaging information around the content. Affiliate programs and subscriptions benefit all parties in the value exchange. The host of the affiliate link or syndicated content provides site visitors with a more diverse experience. (In addition, the host gains revenue from transactions consummated through an affiliate link.) The content creator or merchant extends the reach of his or her brand. Syndication content providers also collect a licensing fee for the use of their content. And the site visitor benefits from a richer experience. However, neither affiliate programs nor syndication benefits publishers unless the relationship is automated and scalable.

Affiliate Program Management

An inefficient affiliate management program reduces the number of affiliate partners a publisher can support and complicates the management of existing relationships. Once an affiliate relationship has been established, ongoing management requires the following:

- *Enrollment:* qualifying and entering affiliate information into your tracking system
- *Tracking:* collecting data on click-throughs and transaction rates per affiliate
- *Remittance:* calculating and remitting referral rewards to affiliates

Unless affiliate management is the site's core competency, using a publishing system that supports affiliate relationships (such as Allaire's Spectra) or outsourcing to a service bureau can save time, effort, and headaches. Outsourcing affiliate relationship management is a benefit to Web publishers who have a large number of affiliates. BeFree, an affiliate program service bureau, handles affiliate relationships from setup through the lifetime of the exchange. Many BeFree customers, such as Barnes&Noble, use commerce-oriented Web sites that lack the tools to manage affiliate programs. Barnes&Noble

(*www.bn.com*) promotes its affiliate program, which contains more than 120,000 participants, with both online merchants and Net users (Guglielmo, 1999).[4] The positive response to the program lies in the relatively high bounties offered by the online merchant. The bn.com site gives affiliate partners a 5% to 7% cut of every transaction made by a Net user referred to the site—for the purchase made through the initial referral and every future purchase by the Net user. In addition, bn.com brings in additional revenue through Net user referrals for products. Barnes&Noble's e-mail referral program allows Net users to forward purchase links in e-mail that connect to the bn.com Web site. Users get a 5% bounty fee for the purchase. Net users collect their reward quarterly in check form delivered by postal mail. In a nice touch, the bn.com e-mail referral program also allows Net user affiliates to donate their proceeds to a charity. Barnes&Noble manages the multiple programs through the affiliate program service bureau BeFree. BeFree handles all the enrollment, tracking, and remittance rewards for bn.com's merchant and Net user affiliates.

Avoiding Affiliate Fraud

Certain pricing structures, such as set fees for click-throughs, are particularly vulnerable to fraud. For example, an unscrupulous affiliate could set up your link on its site and run a script that continuously clicks on the link. Since a set fee-payment structure awards affiliates by click-through, the merchant pays for the script-generated activity. To prevent affiliate fraud, set up reward structures based on commerce or membership. In these scenarios, affiliates get paid when they refer a Net user to a Web site, who then makes a purchase or registers as a member of the site community. That way, the merchant rewards affiliates only for a proven benefit from the Web site.

Web sites with affiliates in the dozens or even hundreds rather than thousands can implement their own affiliate support program through Web publishing systems rather than outsourcing or buying special software to manage the relationships. For example, the Spectra Web publishing system allows site affiliates to set up and manage relationships with a site publisher through a

---

[4]*www.bn.com* is the online counterpart of bookseller Barnes&Noble, Inc.'s more than 1,000 bricks-and-mortar stores. The bn.com site was launched in May 1997 and is currently one of the top five largest e-commerce retailers on the Web with more than 750,000 book, music, and magazine titles.

Web-based interface. Building affiliate infrastructure in house requires extensive resources to manage the tracking and billing. Large-scale affiliation management is best outsourced to an online service bureau, such as BeFree, or through specialized software.

Syndication Management

Like affiliate programs, syndication requires a highly automated delivery and management process. Publishers use syndication to increase traffic and revenue at a Web site. Subscriptions and transactions require very different processes for content exchange. As a syndicate partner, you need access to certain types of data such as click-through rates and rankings of content by popularity. From a business perspective, you monitor syndication relationships for the return on investment for the partnership. Key performance indicators for syndicated content include the number of page impressions or purchases. From a technical perspective, you also need to provide a channel to allow your partners to send new syndicated content quickly and easily to your Web site.

Increasing Traffic through Syndication

With syndication, a publisher adds interest through diversity of content. Yahoo! distributes content from hundreds of providers, ranging from news services to teen magazines. Start-up fitness portal FitForAll provides Net users with syndicated content from health magazines and vitamin retailers. Fidelity uses syndicated TheStreet.com content for its intranet. Syndication aimed at increasing traffic generally uses a subscription model in which the host content partner pays the content provider in defined time increments (usually monthly).

Syndication Roles

From the distributor's perspective, syndication requires several stages. First, the distributor must be able to easily select and review syndication options from content providers. Second, the distributor integrates the in-bound syndicated content into the existing Web site. Even though the content is provided by another publisher, the distributor presents the information in a manner that doesn't disrupt the site's design and user flow. Finally, the distributor tracks the usage of syndicated content to determine which types rank the most popular among site visitors. Publishers build their own systems to manage these steps, buy software solutions such as Kinecta's Interact system, or outsource the relationship to a service bureau such as iSyndicate.

Content providers also require automated support to be able to offer and syndicate content to subscriber sites easily. Content providers syndicate content to distributors via a service bureau or software that supports the XML-based document delivery. Emerging protocols, such as Internet Content Exchange (ICE) and Web Distributed Data Exchange (WDDX), facilitate the exchange of content by standardizing the format for delivery between content providers and distributors. In addition, content providers manage and bill distributor subscriptions.

Content publishing systems provide software that automates the syndication of information between content partners and distributors. Vignette's Syndication Server operates with StoryServer to manage and syndicate content using the ICE protocol. ICE provides a standard for servers to exchange content packaged in XML. Kinecta's Interact software provides content providers and distributors with a low-cost alternative to syndication management, without the workflow management features of a Vignette solution. Allaire's Spectra uses a syndication component, with data transferring via WDDX. WDDX takes the notion of syndication to the application level and offers an XML-based technology that enables the exchange of complex data between Web programming languages.[5] Allaire's syndication model also exposes an XML-based object layer that developers use to automate remote processes in any application language. Regardless of the system chosen, syndication relationships—like affiliate programs—require a high degree of automation to scale to the many relationships a single publisher may manage.

---

[5] WDDX consists of a language-independent representation of data and a set of modules for programming languages, including Cold Fusion, ASP, COM, JavaScript, Perl, Java, PHP3, and Python.

Strategy Summary: Develop Scalable Affiliate and Syndication Management Programs

Business Drivers: Improved productivity, better tracking of syndication and affiliate relationship results

Value Exchanges: Content, commerce

Site Impact: Impacts areas on the site that syndicate content and areas in which affiliate links are offered

Content Business Models: Syndication, affiliate commerce

Best Examples: Amazon (*www.amazon.com*), Barnes&Noble (*www.bn.com*), CDNow (*www.cdnow.com*), Reuters (*www.reuters.com*), TheStreet.com (*www.thestreet.com*), Yahoo! (*www.yahoo.com*)

Summary

The workflow and internal processes discussed in this final chapter illustrate requirements for supporting killer content, such as differentiating the needs of a commerce versus content publishing system, managing effective workflows, using business intelligence, and scaling syndication of affiliate relationships. Automation supports the ability to continuously update content and services for consumers. By automating manual processes, publishers reduce the time to market and level of human error for Web site updates, affiliate management, syndication management, and content business management. These processes provide the backbone for value exchange on your Web site.

The strategies in this chapter, like all the strategies in Part II, provide examples of how to implement value exchange on your Web site. These strategies present the benefits and risks of the mechanisms for building killer content and services on your Web site that were identified in Part I. It would be impossible to describe all of the strategies for improving a Web site here, so we have focused on items that are practical and have significant impact on the relationship with the customer. These items constitute the killer content and services that help formulate a sustainable relationship with a consumer. As mentioned before, there's no single cookie-cutter approach to setting up great services for value exchange on your Web site, just as there's no single business model that fits all Web businesses.

The value—and the fun—of building out your own premium content and services culminates in the relationship that you establish with site visitors and customers. Killer content and its supporting processes give a publisher the ability to either stand out from the crowd or fade into the noise on the Internet.

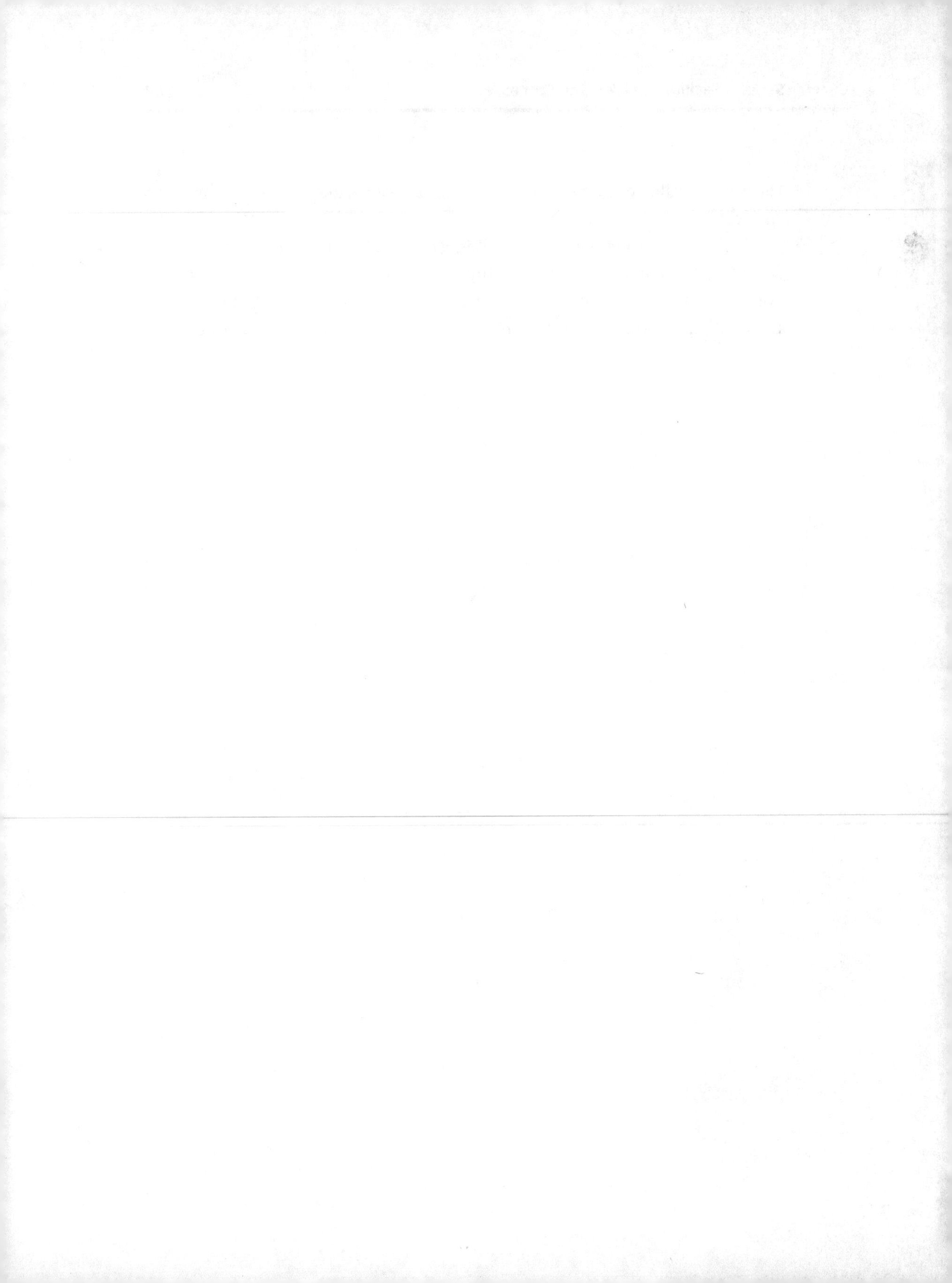

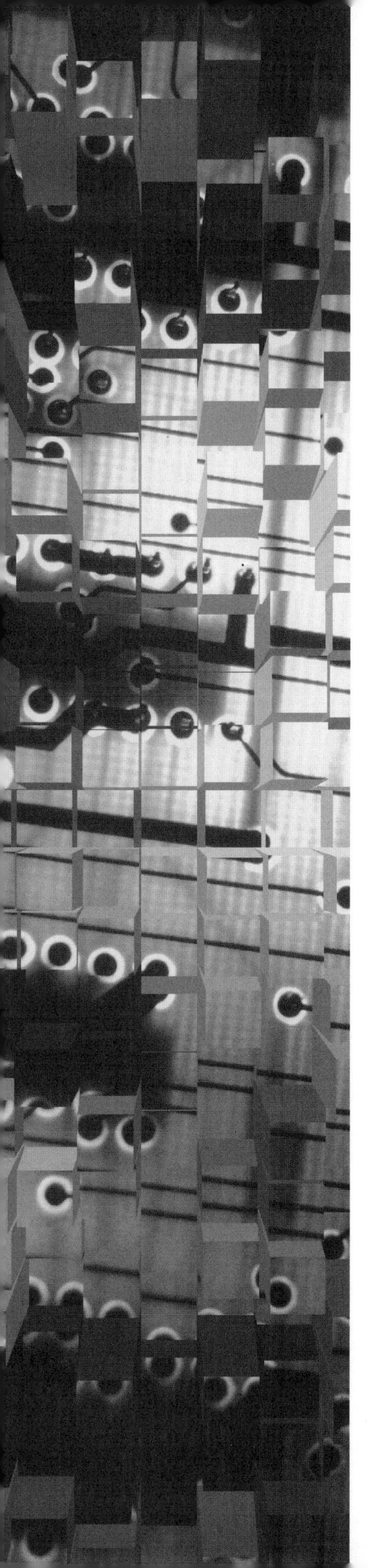

Epilogue

The boundaries between content and commerce blur as the Web evolves into that "realistic mirror"—reflecting how we work, play, and live—that Tim Berners-Lee predicted. A commerce transaction might hinge on syndicated content or offer a Webcast of a rock star's concert. Ignoring Net users' needs and expectations around killer content and services greatly diminishes a publisher's ability to build a sustained revenue model.

A Web publisher is in the business of providing maximum services and value to customers, whether those customers are Net consumers or other publishers. A single-minded focus on one aspect of the value exchange, such as cost, lowers the barrier to entry for competitors, who may steal market share in the publisher's space.

The most exciting developments in killer content are the new pricing and distribution models. Pricing models that support the merchandizing of content allow publishers to turn content into a product that can be cataloged, inventoried, and sold. That fundamental shift in the nature of content opens up an entirely new arena for content usage on the Internet, including pay-as-you-go fees for software on the Internet and other mechanisms. In addition, the ability to bundle multiple items of content into a single item of greater value allows content publishers to enter the world of retail and take part in e-commerce.

Delivery mechanisms such as syndication have been around in the print world for decades but seem made especially for the Internet. With the advent of protocols that facilitate the transfer of information between sites and applications, syndication of content will loosen the strings on content location. The economics of demand will take over, and killer content on the Web will be syndicated to multiple locations, with licensing fees claimed by the content providers. A truer evolution will occur when content moves freely from the Internet to television and to consumer platforms such as handheld computers and cellular phones. Imagine a free flow of information and commerce between the Internet and consumer devices such as your television, phone, car, pager, handheld computer, or smart card. Even with the increasing flexibility of the pricing models and distribution mechanisms, a simple fact remains: killer content lies at the core of the value exchange with consumers.

Glossary

ActiveX Component-based technology downloaded from a Web page to a user's computer in order to enable additional functionality at a Web site.

Ad inventory Advertising slots for a Web publisher's site.

Advertising The practice of advancing a company's product or brand on the Internet through banners or other graphics on a Web page. As a revenue stream, price is often set by site traffic and demographics.

Advertorial In print media, an editorial, article, or opinion that also has an advertising component. In online media, an editorial, article, or opinion that has an interactive or e-commerce capability built into it.

Affiliate manager The Web merchant providing the products or content for an affiliate program.

Affiliate partner A Web publisher who posts a link to products or content at another Web site or merchant.

Affiliate program A mechanism by which Web merchants allow other Web publishers to post links for the store's products or content on different Web sites. The Web merchant provides some benefit, generally a percentage of any purchases made through the link, to the Web site partner.

Applet A Java program that runs in the context of a Java-capable browser or the applet viewer. Java applets extend the content of Web pages beyond just graphics and text.

Auction pricing Pricing model whereby buyers and sellers negotiate an acceptable value for a product through bids.

Banner Graphic on a Web page that generally displays advertising for a company.

Beenz Internet currency that acts as a gift certificate. A consumer obtains Beenz that can be redeemed for value on Web sites that accept them. Also the name of the company that manages the award and redemption of the units.

Broadband access Dedicated high-speed Internet access via cable, wireless, satellite, or existing telephone lines. This technology provides access to content from low-integrity media (such as text and tinny-quality sound files) to high-quality, high-resolution media (such as MP3 files and streaming video).

Broadcasting The ability to distribute live television or events on the Internet via streaming audio and video technologies. Also referred to as Webcasting.

Business analysis Reports on Web site usage based on site traffic patterns and trends, including any transactional information and user response to marketing programs.

Business intelligence Information about the state of a Web-based business, ranging from purchase patterns to Web site reliability.

Business rules Logical rules that drive the business and service model for the Web site. For example, if a person files a legitimate grievance through customer service, business rules dictate that the customer receives a $10 gift certificate.

Business-to-business The sale of product and services from one Web merchant or publisher to another.

Business-to-consumer The sale of product and services from a Web merchant or publisher directly to a Net consumer.

Button Graphic on a Web page.

ClickMiles Individual unit redeemable for frequent flyer miles in the Netcentives ClickRewards program. Used to encourage loyalty and for other promotional campaigns for Web sites.

Click through The process of clicking on a link on a Web page.

Click-through A metric indicating that the link has been clicked on the Web page.

Click-through rate The number of times that Net users clicked on an advertisement relative to the number of times the Web page was displayed to Net users. (Assumes that the banner advertisement provides a link.)

Collaborative filtering The analysis of purchase patterns and preferences to match up a Net user with other Net users with similar buying habits, and to make recommendations based on the match.

Commerce value exchange The exchange of value between a Web merchant offering products and a Net user offering payment. Commerce products include physical goods, such as books, as well as digital goods, such as information.

Commodity content Content that is freely and widely available on the Web.

Consumer merchandizing Commerce conducted by Net users selling goods and services to other Net users.

Content Text, images, audio, and video that compose a Web site.

Content provider Web publisher who creates original content.

Content value exchange The exchange of value between a Web publisher who offers information and Net users who create site traffic to drive advertising.

Conversion Transformation of a casual visitor into a frequent visitor, a visitor into a first-time buyer, or a first-time buyer into a repeat customer.

Conversion rate The rate at which the Net user actually performed the target objective of a link. Generally calculated by the number of users who successfully completed the process divided by the total number of people who clicked the link or viewed the Web page. Typically, conversion rates describe completed purchases and registrations from a banner advertisement.

Cookie Small text file stored on a Net user's machine.

CPM Cost per thousand. Web site page views are aggregated into the thousands in order to effectively describe the rate at which advertisements are charged.

Credibility Value factor that emphasizes the reliability and authority of the source of content on a Web site.

Cross-sell A marketing tactic to offer similar products in order to encourage the Net user to buy more.

Dial-up Connections to the Internet made through phone-based modems.

DigiBox Proprietary file format developed by InterTrust Corporation for digital rights management.

Digital goods Content or software distributed through the Internet. For example, the latest upgrade of a software program downloaded from the Web, or an MP3 song file.

Digital image providers Publishers who digitize images and stock photography and provide the online images to Net users and businesses, usually for a fee.

Digital rights management (DRM) Managing access to content distributed via the Internet, based on the content's copyright and usage restrictions.

Digital subscriber line (DSL) Dedicated high-speed Internet access over existing telephone lines.

Digital watermarking A rights management technique that focuses on identifying the origin and owner of the content in order to discourage and identify unauthorized usage. A digital watermark is a set of data that is embedded in an audio or video file. First adopted by digital image providers such as Corbis Images, it is considered one solution for digital image and music distribution.

Direct e-mail marketing Marketing campaigns conducted using e-mail.

Distributor Web site that provides the syndicated content to Web users.

Document Type Definition (DTD) The rules for document construction data exchange via XML; it defines the nature of every element in an XML file.

Electronic Commerce Modeling Language (ECML) Proposed standard with the World Wide Web Consortium (W3C) for forms used in e-commerce, such as shipping address forms.

Entertainment value exchange The exchange of value between a Web merchant who provides entertaining content and Net users who provide loyalty and site traffic to drive advertising or cross-site referrals.

eXtensible Markup Language (XML) Standard for a markup language (like HTML) that allows users to store and exchange data in a structured format over the Internet.

Fantasy auctions Online auctions that combine fictional commodities and real events to drive an artificial bid-and-exchange.

Fantasy sports Statistics-based competition in which Net users can build virtual teams based on the statistics for real-life athletics.

Free content value exchange Exchange of Net browser loyalty and patronage for free, continuously updated content and services.

Gift certificate currency Unit of currency that provides Net users with the ability to give other Net users a non-merchant-specific gift certificate.

Hard goods Various physical products sold over the Internet and distributed via supply–distribution chains.

HyperText Transfer Protocol (HTTP) Standard transmission protocol for the World Wide Web.

Information Content Exchange (ICE) Proposed protocol for the automatic, controlled exchange and management of online assets directly between Web servers over the Internet.

Information structure The layout and organization of information on a Web site that enables a consumer to take advantage of content and services.

Innovation Value factor that emphasizes the uniqueness of the content.

Intelligent agent (bot) Software program that performs an action on a Net user's behalf.

Interstitial An online advertisement that opens a separate browser window when a Web page loads.

ISDN Integrated Services Digital Network, which is a digital service operating over existing copper telephone wiring.

"Killer app" A concept or application that enacts radical change in the online world.

Killer content High-quality services and content offered in exchange for Web site loyalty and commerce.

Media cannibalism Diluting the pool of purchasers from one form of revenue-generating media (e.g., print newspaper) by offering the same content for free through another medium (e.g., on-line newspaper).

Membership The state of belonging to an organization or association based on registration or some other affiliation; the number of Net visitors who have registered at a Web site.

Merchandizing content The packaging of content (text, images, and rich media) into a product that can be bought and sold by Net users and businesses.

Monetize To create revenue streams from content and services.

MP3 file format MPEG-2 Audio Layer-3. Compresses a sound file to one-twelfth its original size, with nearly CD-quality sound.

MP3 player Software that decompresses MP3 files and essentially "plays" the song.

Net browser Net user who travels to one or more destinations on the Internet to read, research, e-mail, and perform other activities that do not require a payment.

Net buyer Net user who makes a purchase through one or more Web sites. Purchased goods range from actual commodities (e.g., books) to premium content (e.g., subscription access).

Net user Someone who uses the Internet for Web browsing or purchasing.

Offline Actions that occur off the Internet. Conversely, the term *online* describes items on the Internet.

Online Actions that occur on the Internet. Conversely, the term *offline* describes items in the physical world.

Online escrow Funds or goods delivered by one person to another which, on the fulfillment of a certain condition, is delivered to a third party. For an online auction, funds can be held "in escrow" until the buyer gets and approves the merchandise, after which the funds are released to the seller.

Online services Services offered to help Net users perform an action, such as holding payment in escrow for a buyer and seller in an online auction or comparing prices automatically on the Web.

Opt-in Describes the status of a member or Net user who explicitly agrees that a Web site can share visitor information under defined conditions.

Page hits The number of times a Web browser requests a file.

Page impressions The number of times a Web file is displayed to a Net user.

Page views The number of times a Web page, with all its associated files, is viewed by Net users.

Pay per access Price set for access to a single product or download, on a per-item basis. Also referred to as "à la carte" products or pay per item.

Personalization The technique of displaying certain Web pages based on Net user profiles or known behavior.

Portal Web site that aggregates a wide variety of content, services, and resources in one area for users.

Premium content Content and services that are not widely available and that have high enough value for consumers to pay for access.

Premium content value exchange Exchange of Net buyer's payment or personal information for high-value content or services.

Price point Price set for content or product by the web publisher.

Privacy policy Policy that describes what a Web publisher plans to do with Net user information obtained through registration or browsing on the Web site.

Privacy seals Approval marks (usually graphics) that denote a Web site's compliance with a privacy program.

Production monitoring Reports on the usage and status of servers supporting Web sites and site applications, including real-time monitoring and analysis of historical trends for operations.

Promotional currency Nonlegal tender awarded to consumers for completing an action on a Web site, which can be redeemed for purchases or other benefits.

Promotional e-mail marketing Marketing campaigns used to promote products and services over e-mail to a list of members who have implicitly or explicitly agreed to receive promotional e-mails from the vendor running the campaign.

Promotional value exchange Limited value exchange whereby a Web publisher provides single-topic information about a particular product or company in exchange for brand recognition.

Prospective e-mail marketing Marketing campaigns used to promote products and services over e-mail to a general list of Net users.

Pull personalization Personalization services for Net users who find information by browsing a personalized Web page.

Push personalization Personalization services for Net users who set preferences online and receive personalized content delivered via e-mail or to another Web site.

Registration The act of providing information in exchange for membership.

Relevance Value factor that depends on the importance and impact of content and services on a decision, goal, or lifestyle.

Reward currency Unit of exchange that allows Web publishers to provide value to Net consumers.

Rip Transferring songs on music CDs to MP3 format.

Satellite connection Form of dedicated high-speed Internet access that requires a PC satellite dish, an unblocked view of the southern sky, and a regular phone line for uploads.

Secure distribution A rights management technique that requires custom software to control and track consumer access to premium or high-value content. Primarily this benefits business-to-business transfer of extremely valuable information.

Secure Encryption Technology (SET) E-commerce security protocol that ensures the cardholder, the merchant, and the acquirer can be fully authenticated.

Services Online tools to help Net users perform an action or task. The action can be holding payment in escrow for a buyer and seller in an online auction, or automated price comparisons on the Web.

Shopping network Portal or other aggregating Web site that offers links to a set of shopping Web sites.

Site features Features and processes that support content and commerce exchange.

Spam E-mail advertisements and promotions sent indiscriminately to a large audience of recipients.

Spider Search engine component that visits Web pages and follows links.

Splash page Temporary graphics displayed to Net users when they open an application or arrive at a Web site.

Sponsorship Advertising relationship that allows a Web publisher to obtain long-term payment in exchange for prominent text and branding

Steganograph Document watermark that is invisible to the eye but can be viewed by special watermarking devices.

Streaming audio Software solution that allows Net users to listen to a digitally encoded audio file on a remote server without requiring the file to be downloaded to their local hard drive.

Streaming video Software solution that allows Net users to see a digitally encoded video file on a remote server without requiring the file to be downloaded to their local drive.

Subscription Price set according to time-based access to content. Can apply to an entire Web site or some subsection of the Web site.

Superstitial Advertisement that loads into a separate browser window *after* the browser downloads and renders the Web page.

Syndication Distribution of content via multiple channels online or offline.

Syndicator Software or online service bureau that aggregates syndication relationships and distributes content for the content provider.

Timeliness The immediacy of content or product, and its relevance for time-sensitive business decisions.

Up-sell Marketing tactic to promote a product of higher value to a Net user who has bought a similar product on a Web site.

Utility The usefulness of the content or product to the Net user's goals and objectives.

Value exchange The exchange of content and services from a Web publisher in exchange for payment or loyalty from the Net user.

Vortal A vertical portal, or a Web site, that aggregates content, services, and resources based on a specific subject or theme.

Webcast Broadcast of live television or radio programming over the Internet.

Web Distributed Data Exchange (WDDX) An XML-based technology that enables the exchange of complex data between Web programming languages.

Web publishers Those who are involved with developing and producing a Web site, whether a senior manager, a Web developer, or a copywriter.

Webzine An online magazine, usually focusing on editorial/commentary content.

Workflow The process of generating and posting content and services to a Web site.

World Wide Web Consortium (W3C) Standards body that regulates protocols and standards used on the Web.

Bibliography

Allen, Lisa. "iSyndicate Solves Content Headaches." Boston: Forrester Research, Inc., 29 Mar. 1999.

Anderson, Diane. "Web Ads: Not Just for Banners Anymore." *The Industry Standard*, 15 Oct. 1999.

Berners-Lee, Tim. "W3C and Standards," Frequently Asked Questions by the Press (*www.w3.org/People/Berners-Lee/FAQ.html*).

Berners-Lee, Tim. "The World Wide Web: A Very Short Personal History." W3C (*www.w3.org/People/Berners-Lee/ShortHistory.html*), 7 May 1998.

Byrne, Tom. "BID & ASK: Where Are Priceline.com's Bulls Now?" Individual-Investor.com, 16 Sept. 1999.

Charron, Chris. "The Content–Commerce Collision." Boston: Forrester Research, Inc., March 1999.

Clemmer, Kenneth. "Media Cannibalization: Myth and Reality." Boston: Forrester Research, Inc., 28 Dec. 1998.

Conlin, Robert. "Microsoft Slams Priceline Suit as 'Desperate Attempt.'" *E-Commerce Times*, 14 Oct. 1999.

Cox, Beth. "Study: Consumers Want More Online Incentives." *E-Commerce News* (*www.internetnews.com*), 28 June 1999.

Daly, James. "The Art of the Deal." *Business 2.0* (*www.business2.com*), Oct. 1999: 106–116.

de Jonge, Peter. "Riding the Wild, Perilous Waters of Amazon.com." *New York Times Magazine*, 14 Mar. 1999.

Delhagen Kate, Emily Nagle Green, Sarah Gerber, and Shar Van Boskirk. "The Privacy Bomb." Boston: Forrester Research, Inc., August 1997.

eMarketer (*www.emarketer.com*) Web site articles: "Web Ad Spending (Projected)," "Advertising Revenues: Ad Dollar Concentration," "Ecommerce: Corporate Experience with Websites," "Net User Demographics: Income," "Paid Content on the Net," "Usage Patterns: Navigational Tools," "Usage Patterns: Reasons for Being Online," and "Web Advertising Surging Ahead," 1999.

Evans, Jim. "The Wide Web of Sports." *The Industry Standard,* 26 July 1999.

Foster, Cormac. "Customer Service Improving, But Sites Still Have a Long Way to Go." New York: Jupiter Communications, 12 July 1999.

France, Mike, Timothy J. Mullaney, and Diane Brady. "Priceline: A Net Monopoly No Longer." *BusinessWeek ONLINE,* 27 Sept. 1999.

Frauenfelder, Mark. "The Future Is at Hand." *The Industry Standard,* 9 July 1999.

Frost, Robert. Edward Connery Lathem (Ed.), "The Aim Was Song." In *The Poetry of Robert Frost.* New York: Henry Holt & Co., Inc., 1979.

Gately, Gary. "Report Finds Top Sites Control Most Ad Revenue." *E-Commerce Times,* 21 June 1999.

Goode, Erica. "The Online Consumer?: Tough, Impatient, and Gone in a Blink." *New York Times,* 22 Sept. 1999.

Godin, Seth. *Permission Marketing: Turning Strangers into Friends, and Friends into Customers.* New York: Simon & Schuster, 1999.

Godin, Seth. *Emarketing.* New York: Berkley Publishing Group, 1995.

Guernsey, Lisa. "Night of the Living Bid: Four Tales from an Hour of eBay." *New York Times,* 22 Sept. 1999.

Guglielmo, Connie. "BarnesandNoble.com to Pay for Referrals." *Internet Week,* 12 July 1999.

Hagen, Paul R. "Streamlining Web Payments." Boston: Forrester Research, Inc., April 1999.

Hardie, Mark E. "Hooked on Broadband." Boston: Forrester Research, Inc., July 1999.

Haylock, Christine Ford, and Len Muscarella. *Net Success: 24 Leaders in Web Commerce Show You How to Put the Web to Work for Your Business.* Holbrook, MA: Adams Media Corporation, 1999.

Hellweg, Eric. "MP3.com: Now What?" *Business 2.0,* Nov. 1999: 87–92.

Hoover's Online. Amazon.com., Digimarc Corporation, eBay, Inc., Priceline.com, Quokka, Razorfish, Inc., Salon.com, TheStreet.com, and Yahoo!, Inc. Company Capsule, Financial, and Profile. Austin, TX, 1999.

Ibanez, Ardith, and Natalie Zee. *HTML Artistry: More Than Code.* Indianapolis: Hayden Books, a division of Macmillan Computer Publishing, 1998.

The Industry Standard. "Hey!!! Webvan's in a Quiet Period!!!" 7 Oct. 1999.

jupdata. "Email Response Rates," Jupiter Communications Online Intelligence. New York: Jupiter Communications, 1999.

jupdata. "Suggestive Selling Online." New York: Jupiter Communications.

Jupiter Communications Press Release. "Kids and Teens to Spend $1.3 Billion Online in 2002." New York: Jupiter Communications, 7 June 1999.

Jupiter/NFO Consumer Survey. "Attitudes, Behaviors, and Demographics of the Online User." New York: Jupiter Communications, 1998, 1999.

jupstat. "Looking for Loyalty." New York: Jupiter Communications.

jupstat. "Unsolicited Email." New York: Jupiter Communications.

jupstat. "Willingness to Register." New York: Jupiter Communications.

Kalakota, Ravi, and Marcia Robinson. *E-Business: Roadmap for Success.* Reading, MA: Addison Wesley Longman, 1999.

Keane, Patrick. "The Sports Portal: ESPN.com Relaunches as Sports Content Aggregator." New York: Jupiter Communications, 8 Sept. 1998.

Lasica, J.D. "The Confidence Game." *The Industry Standard,* 5 Mar. 1999.

Ledbetter, James. "Your Band Isn't Really That Popular After All." *The Industry Standard,* 13 Sept. 1999.

MacKenzie, Meghann, and Shelley Morrisette. "The Internet Shift Endangers Financial Print Media." Boston: Forrester Research, Inc., 3 Sept. 1998.

MacKenzie, Meghann, Kenneth Clemmer, and Shelley Morrisette. "Consumers Are Ready for Broadband Technologies." Boston: Forrester Research, Inc., 16 Sept. 1998.

Masterson, Michele. "FTC: E-Comm Sites Need to Give Consumers More Information." *E-Commerce News,* 10 June 1999.

McKay, Niall. "eBay's Fortune." Redherring.com, 28 Sept. 1999.

McQuivey, James L. "Web Travel Winners." Boston: Forrester Research, Inc., June 1999.

Mooradian, Mark, Regina Joseph, and Patrick Keane. "Navigation: Towards Intuitive Movement and Improved Usability." New York: Jupiter Communications, March 1999.

Mooradian, Mark. "Media Cannibalization: Radio and Newspapers Have Most to Fear." New York: Jupiter Communications, 23 Nov. 1998.

Morrisette, Shelley. "Are Net Shoppers Loyal?" Boston: Forrester Research, Inc., March 1999.

Nail, Jim. "Driving Site Traffic." Boston: Forrester Research, Inc., April 1999.

O'Hara, Colleen. "Vendor to Offer Training Via Internet." *Federal Computer Week*, 22 Feb. 1999.

Online Computer Library Center, Inc. (OCLC), Office of Research. "Internet Growth Statistics" (*www.deepcanyon.com/b/numbers/nn10041999.htm*), 4 Oct. 1999.

Reamer, Scott. "The Internet Capitalist" newsletter. SG Cowen, 23 Apr. 1999, 3 Aug. 1999, 24 Sept. 1999.

Richtel, Matt. "Man Emerges, Still Breathing, After E-Cave Confinement." *New York Times*, 22 Sept. 1999.

Sacharow, Anya. "GO's Entry Reaffirms Necessity of Promotion and Media Partners." New York: Jupiter Communications, 21 Dec. 1998.

Schoenfeld, Adam. "Bloomberg, CNET Get Personal on the Web." New York: Jupiter Communications, 9 Feb. 1998.

Schwartz, Evan I. *Digital Darwinism: 7 Breakthrough Strategies for Surviving the Cutthroat Web Economy*. New York: Broadway Books, 1999.

Shore, Melissa, Fiona S. Swerdlow, Heidi Kim, and Marc Johnson. "Travel: Online Landscape." New York: Jupiter Communications, October 1999.

Shuchman, Lisa. "Online Travel Takes Off." *The Industry Standard*, 8 Oct. 1999.

Siegel, David. *Creating Killer Web Sites: The Art of Third-Generation Site Design, Second Edition*. Indianapolis: Hayden Books, a division of Macmillan Computer Publishing, 1997.

Sinnreich, Aram, Marc Johnson, Mark Mooradian, and Anya Sacharow. "Copyright and Intellectual Property: Creating New Business Models with Digital Rights Management." New York: Jupiter Communications, June 1999.

Speigel, Rob. "Amazon.com and American Airlines Top Internet Customer Loyalty List." *E-Commerce Times*, 13 Oct. 1999.

Sterling, Robert, Marc Johnson, and Robert Leathern. "Financial Services Projections." New York: Jupiter Communications, August 1999.

Tedeschi, Bob. "CDNow Struggles to Be Heard." *New York Times*, 24 May 1999.

Tedeschi, Bob. "Ecommerce Report: Good Web Site Design Can Lead to Healthy Sales." *New York Times*, 30 Aug. 1999.

Thomas, Owen. "ICE May Unstick Content Markets." *Red Herring: The Business of Techology* (online), 11 Feb. 1998.

Warner, Bernhard. "Razorfish Catches Another One." *The Industry Standard,* 10 Aug. 1999.

Werbach, Kevin. "The Web Goes into Syndication." *Release 1.0, Esther Dyson's Monthly Report* (*www.edventure.com*), 27 July 1999.

Williamson, Debra Aho. "Virtual Vineyards Pumps Bottom Line with Email." *Advertising Age,* 19 Apr. 1999.

Wolverton, Troy. "The Costs of E-commerce." CNET News.com, 28 May 1999.

Index